SpringerBriefs in Environmental Science

SpringerBriefs in Environmental Science present concise summaries of cutting-edge research and practical applications across a wide spectrum of environmental fields, with fast turnaround time to publication. Featuring compact volumes of 50 to 125 pages, the series covers a range of content from professional to academic. Monographs of new material are considered for the SpringerBriefs in Environmental Science series.

Typical topics might include: a timely report of state-of-the-art analytical techniques, a bridge between new research results, as published in journal articles and a contextual literature review, a snapshot of a hot or emerging topic, an in-depth case study or technical example, a presentation of core concepts that students must understand in order to make independent contributions, best practices or protocols to be followed, a series of short case studies/debates highlighting a specific angle.

SpringerBriefs in Environmental Science allow authors to present their ideas and readers to absorb them with minimal time investment. Both solicited and unsolicited manuscripts are considered for publication.

Carolina S. Vera • Ileana Bladé
Brant Liebmann

Climate Variability and Change in Southern South America

Focus on the La Plata Basin

Springer

Carolina S. Vera
Centro de Investigaciones del Mar y la Atmósfera (CIMA) – Departamento de Ciencias de la Atmósfera y los Océanos (DCAO)
Facultad de Ciencias Exactas y Naturales, Universidad de Buenos Aires (UBA)
Buenos Aires, Buenos Aires, Argentina

Ileana Bladé
Grup de Meteorologia, Departament de Física Aplicada, Facultat de Física
Universitat de Barcelona
Barcelona, Barcelona, Spain

Brant Liebmann
San Diego, CA, USA

ISSN 2191-5547 ISSN 2191-5555 (electronic)
SpringerBriefs in Environmental Science
ISBN 978-3-032-07196-5 ISBN 978-3-032-07197-2 (eBook)
https://doi.org/10.1007/978-3-032-07197-2

This Springer imprint is published by the registered company Springer Nature Switzerland AG
The registered company address is: Gewerbestrasse 11, 6330 Cham, Switzerland

Preface

At the beginning of the twenty-first century, and over the following decade, the scientific communities from southern South American countries—including Argentina, Brazil, Chile, and Uruguay—forged partnerships with research groups from the United States and Europe around shared interests in the region's climate. These collaborations spawned a series of coordinated projects and programs aimed at expanding and deepening the understanding of climate variability in South America. Many of these initiatives were developed under the auspices of the World Climate Research Programme, which helped attract the attention of and facilitated support from major international funding agencies, while also providing structure and resources for scientific projects funded by regional governments.

A large part of this research activity focused on South America's major river basins: the Amazon and, most notably for the present work, the La Plata Basin. These efforts led to substantial progress in understanding the mechanisms driving climate variability in the basin, revealing the key role of interactions between the South American Monsoon System and extratropical variability associated with Southern Hemisphere circulation patterns. In addition, significant progress has been made in understanding how the regional climate is modulated by large-scale phenomena such as the El Niño–Southern Oscillation (ENSO) and the Southern Annular Mode (SAM), as well as by long-term anthropogenic climate change.

This expanded scientific understanding has, in recent decades, supported the development of better climate monitoring systems and enhanced the ability to issue forecasts weeks to months in advance. Such improved tools are especially valuable for informing decision-making in climate-sensitive sectors within the basin, such as agriculture, energy, and tourism. For this reason, it is increasingly important that knowledge about regional climate variability and change be made accessible to professionals and experts beyond the field of climate science.

This book was born from that motivation and the authors' combined expertise, particularly Carolina's extensive experience teaching courses in Argentina and abroad on climate variability and change in southern South America, with a special focus on the La Plata Basin. Here, we present the most relevant features of regional climate variability and change, using conceptual models to make the material accessible to readers without a prior background in climate science. We also explore current capabilities in climate prediction and highlight ways in which this knowledge can be shared and leveraged through climate services.

Buenos Aires, Buenos Aires, Argentina — Carolina S. Vera
Barcelona, Barcelona, Spain — Ileana Bladé
San Diego, CA, USA — Brant Liebmann

Competing Interests The authors have no competing interests to declare that are relevant to the content of this manuscript.

Contents

Abbreviations

AMO	Atlantic Multidecadal Oscillation
CEMADEN	National Early Warning and Monitoring Center for Natural Disasters
CPC	Climate Prediction Center
CHFP	Climate Historical Forecast Project
CHIRPS	Climate Hazards Group InfraRed Precipitation with Station Data
CID	Climate Impact Driver
CO_2	Carbon dioxide
CPTEC	Centro de Previsão de Tempo e Estudos Climáticos
CRC-SAS	Regional Climate Center for Southern South America
DJF	December–January–February
DSS	Decision support systems
ECMWF	European Centre for Medium-Range Weather Forecasts
EN3.4	El Niño 3.4
ENSO	El Niño Southern Oscillation
EOF	Empirical Orthogonal Function
GFCS	Global Framework for Climate Services
GHG	Greenhouse gases
GPCs	Global Producing Centers
IBF	Impact-based forecasts
IOD	Indian Ocean Dipole
IPCC	Intergovernmental Panel on Climate Change
ISI	Intraseasonal to interannual
ISO	Intraseasonal oscillation
ISV	Intraseasonal variability
ITCZ	Intertropical convergence zones
JJA	June–July–August
LPB	La Plata Basin
MAM	March-April-May
MGB-SA	Modelo de Grandes Bacias-South America
MJO	Madden-Julian Oscillation

NCEP-NCAR	National Centers for Environmental Prediction-National Center for Atmospheric Research
NMHS	National Meteorological and Hydrological Services
NMME	North American Multi-Model Ensemble
NOAA	National Oceanic and Atmospheric Administration
OLR	Outgoing longwave radiation
OND	October-November-December
PDO	Pacific Decadal Oscillation
PSA	Pacific-South America
RD	Remote driver
RDDB	Regional Daily Database
S2S	Subseasonal to Seasonal Project
SACZ	South Atlantic Convergence Zone
SALLJ	South American low-level jet
SAM	Southern Annular Mode
SESA	Southeastern South America
SICZ	South Indian Convergence Zone
SIMAGRI	Simulador de Agricultura
SimSEE	Electric Power System Simulator
SIS	Seasonal-Intraseasonal Pattern
SM	Soil moisture
SON	September-October-November
SPCZ	South Pacific Convergence Zone
SST	Sea surface temperature
TAO	Tropical Atmosphere-Ocean
WIGOS	World Integrated Global Observing System
WMO	World Meteorological Organization

List of Figures

Chapter 1
Introduction

The La Plata Basin (LPB) spans an extensive region of South America (Fig. 1.1), covering approximately 3,500,000 km^2 and encompassing five countries: Argentina, Brazil, Bolivia, Paraguay, and Uruguay. It is home to a population of over 100 million people and contributes to more than 70% of the collective Gross Domestic Product (GDP) of these nations (CIC, 2017). As the second largest basin in South America after the Amazon and the fifth largest globally, the LPB stands out as one of the most important hydrological basins in the world, not only due to its vast size but also because of its profound socioeconomic and environmental significance. The basin's population is concentrated in numerous cities and towns, including two megacities—São Paulo and Buenos Aires—that are among the largest urban centers in the world. The LPB supports a wide range of regionally and globally critical socioeconomic activities, such as food production, electricity generation, river navigation, tourism, and more, all of which are highly sensitive to climate variability and change.

The basin is formed by three major rivers—Paraná, Paraguay, and Uruguay—along with their extensive network of tributaries, which ultimately converge into the Río de la Plata. This hydrological system sustains large natural ecosystems, including the Pantanal of the Paraguay River, the Paraná River Delta, and other significant wetlands, as well as diverse ecoregions ranging from humid forests to "cerrados" (a type of tropical savanna), and temperate grasslands or pampas. These ecosystems provide essential beneficial functions such as water regulation, biodiversity conservation and carbon sequestration, but are increasingly threatened by human activities, particularly land-use changes, and climate change. Among the sub-basins, the Paraná is the largest, covering 49% of the total area, followed by the Paraguay (35%) and the Uruguay (16%) (Perez et al., 2021 and references therein).

The climate of the LPB exhibits significant variability across multiple temporal scales, ranging from daily and weekly fluctuations to seasonal, interannual, and

C. S. Vera et al., *Climate Variability and Change in Southern South America*, SpringerBriefs in Environmental Science,
https://doi.org/10.1007/978-3-032-07197-2_1

Fig. 1.1 General Map of the la Plata Basin (Adapted from CIC (2017)). The red box delimits part of the region commonly referred to in the literature as Southeastern South America (SESA), which extends further west to the foothills of the Andes and further south to around the 40°S (not shown)

even decadal cycles. These variations can result in either favorable or adverse climatic conditions, which in turn have profound impacts on the socioeconomic activities and natural ecosystems within the region. Over recent decades, climate science has made considerable progress in understanding the drivers and dynamics of this regional climate variability. Advances have also been achieved in predicting these variations on timescales ranging from weeks to seasons and even years and in projecting the effects of anthropogenic climate change.

The objective of this book is not to provide an exhaustive or detailed description of the LPB's climate. Rather, it aims to present current and updated knowledge on climate variability in this region in a manner that is accessible to professionals who are not experts in climate science but work in technical fields related to socioeconomic sectors within the basin—such as water resources, agriculture, energy, health —or are involved in the research and management of natural assets tied to ecosystems and biodiversity. To this end, the book focuses on describing the influence of both continental and large-scale (hemispheric and global) climate variability on the regional climatic conditions of the LPB. Among the various climate variables, the book places particular emphasis on precipitation and its variability, given its critical role in shaping natural ecosystems and human activities in the region.

The LPB is so vast that it stretches from the southern tropics (20°S) to the mid-latitudes of Argentina (40°S), covering much of South America east of the Andes, thus spanning both tropical and extratropical regions. Its northern sector includes parts of Bolivia, western Paraguay, and southeastern Brazil and is characterized by a tropical climate with abundant summer rains and high levels of humidity. In contrast, its southern portion lies within the region known as Southeastern South America (SESA), which comprises southern Brazil, eastern Paraguay, northern and eastern Argentina, and Uruguay, and is characterized by a subtropical to extratropical climate, with lower rainfall spread throughout the year, lower humidity levels, and more pronounced temperature variations. Figure 1.1 shows the area generally referred to as SESA in the literature. Most studies on regional climate variability and change focus on these two climatic zones that comprise the LPB, often without explicitly referring to the basin itself. Therefore, this broad division into two distinctive climatic zones is commonly adopted in the descriptions and analyses presented throughout the chapters.

The book is organized as follows: Chapter 2 provides an overview of the data and methods used to study climate variability, as well as the fundamental physical concepts underlying global and South American climate dynamics. Chapter 3 introduces sub-seasonal or intraseasonal variability, which operates on timescales between 20 and 70 days and significantly affects regional climate, although its causes remain relatively poorly understood. Chapter 4 explores the influence of large-scale climate patterns on seasonal and interannual scales, such as the El Niño-Southern Oscillation (ENSO) and other patterns originating in the tropical oceans, as well as modes of variability arising in polar regions, such as the Southern Annular Mode (SAM). Chapter 5 describes the characteristics of regional climate variability on decadal scales, its interaction with long-term climate trends associated with anthropogenic climate change, and future regional climate projections. Finally, Chapter 6 introduces the concept of "climate services," focusing on the development, communication, and application of climate information that is valuable or useful for decision-making in socioeconomic sectors sensitive to climate variability and change. This chapter also includes a brief discussion of current predictability and forecasting capabilities of the basin's climate on subseasonal and seasonal timescales, highlighting the potential for improved climate resilience and adaptation strategies in the region.

References

CIC. (2017). *Transboundary diagnostic analysis for the La Plata River Basin - TDA*. Comité Intergubernamental Coordinador de los Países de la Cuenca del Plata (CIC), Organización de los Estados Americanos, OEA. https://cicplata.org/wp-content/uploads/2017/09/transboundary_diagnostic_analysis_for_the_la_plata_river_basin.pdf

Perez, L., Barreiro, M., Etchevers, I., et al. (2021). Centennial hydroclimatic and anthropogenic processes of south East South America modulate interannual and decadal river discharge. *Science of the Total Environment, 781*, 146733. https://doi.org/10.1016/j.scitotenv.2021.146733

Chapter 2
Basic Climate Features

2.1 Data and Methods Used to Describe Climate Variability

The description of climate variability at local, regional and global timescales is essential for its understanding and for developing tools to monitor and predict its conditions. This goal requires access to data with three-dimensional spatial coverage and temporal records that are sufficiently long to apply appropriate mathematical tools and obtain robust results.

The world's meteorological services conduct routine observations of atmospheric variables, which are shared through the World Integrated Global Observing System (WIGOS) of the World Meteorological Organization (WMO) (WMO, 2019). Established in the 1960s, this system was developed to facilitate the exchange and integration of standard atmospheric observations—such as pressure, temperature, wind, humidity, and rainfall— made at different locations around the globe, which are needed to monitor daily atmospheric conditions and forecast their evolution over the coming days. Over the decades, the system has expanded to include information on ocean conditions and other components of the climate system.

Surface weather stations, whether manual or automatic, provide relatively dense coverage over much of the Earth's land surface and constitute the longest instrumental record of atmospheric observations. Ship-borne measurements and buoy networks (such as the Tropical Atmosphere-Ocean (TAO) array installed along the equatorial Pacific) augment the surface stations and provide information about ocean conditions within the subsurface boundary layer. Ship observations are sparse except along shipping lanes and are not continuous in time. Buoys are also sparse, in part because they are expensive to maintain, but provide continuous data at a fixed location.

Radiosondes are devices equipped with sensors carried aloft by balloons at specific locations. Primarily deployed over land, these devices provide vertical profiles

C. S. Vera et al., *Climate Variability and Change in Southern South America*, SpringerBriefs in Environmental Science,
https://doi.org/10.1007/978-3-032-07197-2_2

of various atmospheric variables with high accuracy. Because radiosondes are relatively expensive and non-recoverable, their networks are considerably less dense than those of surface stations. Moreover, these measurements are typically conducted operationally every 6 h at the most and, in many countries, only once or twice daily.

Satellites provide a vast amount of meteorological data. Geostationary satellites are located high above the Earth and continuously monitor the same sector of the planet; they offer good temporal coverage of specific areas and atmospheric layers, albeit at a modest spatial resolution. Orbital satellites are located in the upper atmosphere and provide high spatial information along their trajectories, resulting in observed areas that vary over time. Satellite data are also important for describing sea surface conditions, such as sea surface temperature (SST), and land conditions, such as vegetation and soil moisture. Satellite information has been available since the late 1970s and early 1980s and has been instrumental in describing meteorological and climatic patterns, particularly over oceanic regions in the data-sparse Southern Hemisphere.

Data from these different observing systems vary in quality, time of measurement, and spatial coverage. Several times a day, the WMO global prediction centers combine data from all these systems to describe current atmospheric and oceanic conditions and to generate the initial conditions for numerical models used in daily and medium-term forecasts. Observations first undergo various quality checks, both individually and collectively. Different mathematical methodologies are then applied to integrate or assimilate these observations. Assimilation involves combining observations from various sources, irregularly distributed in space and time, with very short-term forecasts to produce a new set of mathematically and physically consistent data. Processed data are provided in regular three-dimensional grids at specific time intervals. This numerical output is known as "analysis," and the resulting information is referred to as "analyzed data." The term "analysis" originates from the pre-computer era, when meteorologists manually drew fields of meteorological variables on maps, based on available observations.

The stages of compilation, quality control, and analysis of observations, along with short-term forecasting, define a "data assimilation system." These systems were first implemented for weather forecasting in the 1960s and 1970s at several global forecast centers, such as those in the United States, Europe, and Australia. In the 1980s, the analyzed fields became a common tool for studies of climate variability. However, data resulting from operational analysis systems are affected by changes in models, data assimilation techniques and observation systems, among other factors. Additionally, they only include data available in real time. Consequently, these real-time analyzed datasets may reflect variations not in the climate itself but in the methods used to construct them. To address this issue, in the mid-1990s, forecast centers decided to "freeze" their data assimilation systems to produce internally consistent analyzed fields, incorporating more observations and improving assimilation techniques. The resulting fields are known as "reanalysis" to distinguish them from operational analysis.

The most widely used global reanalysis dataset has been the NCEP-NCAR reanalysis, which provides data from 1948 to the present (Kalnay et al., 1996). The European Centre for Medium-Range Weather Forecasts (ECMWF) has also developed a variety of reanalysis datasets, including the recently implemented ERA-5 (Hersbach et al., 2020). The availability of these global reanalysis datasets has significantly advanced our understanding of climate and its variability. However, despite improvements in the mathematical techniques used to generate them, their quality still varies notably across regions, particularly between periods with and without satellite information. This is especially relevant in the Southern Hemisphere, where satellite data are critical for characterizing climatic conditions over its vast oceanic areas.

Regional climate variability studies, such as those conducted in South America, rely on observations from ground-based systems, particularly surface stations, as well as data from remote sensing and reanalysis systems. The choice of data depends on the study's purpose. While in-situ observations allow for precise descriptions of temporal variability at specific locations, reanalysis datasets provide a broader spatio-temporal characterization of climate variability and change across regions.

In central and southern South America, where the LPB is located, observational networks are generally of medium to low density. Surface stations have become more abundant in certain regions with the recent addition of automatic stations. Manual stations with long, high-quality records are relatively scarce and are irregularly distributed. The radiosonde network is also sparse, with limited coverage in both space and time. Fortunately, satellite data and operational or reanalysis systems have improved the characterization of regional climate variability. Nonetheless, meteorological information remains limited in key areas, such as the Andes Mountains, rainforests, and the Patagonian plateau. Therefore, comparing data from multiple sources is essential to obtain robust assessments of regional climate variations.

2.1.1 Means and Perturbations

To describe climate variability, it is necessary to apply various mathematical treatments to the data. Averages are widely used to synthesize information according to the specific focus of interest. Temporal averages—calculated either over the entire available record or over a sufficiently long period —allow for the determination of long-term average or climatological mean conditions, which describe the fundamental behavior of climate.

At least 30 years of data are required to compute these climatological means, as atmospheric variability often follows a 'red noise' pattern, meaning low-frequency (long-period) fluctuations tend to have greater amplitudes. As a result, long-term records are essential to avoid biases from specific phases of these oscillations. Even then, climatological means may not be stationary, particularly in the presence of long-term trends associated with anthropogenic climate change. Seasonal and monthly means are used for examining climate anomalies during specific times of the year, while filtering out shorter-term variations. Studies of interannual variability often rely on monthly-mean data. Quarterly averages are frequently used to describe seasonal conditions (e.g., December–January–February, or DJF). For example, as discussed in Sect. 2.3, the long-term averaged precipitation in South America for DJF and JJA (June–July-August) allow us to describe and distinguish the seasonal differences of this variable in the different regions.

Another commonly used approach is the temporal average of specific data, or "composites", which involves averaging those that meet certain criteria or fall into defined categories. For instance, compositing may be applied to assess the influence of ENSO on seasonal precipitation: to filter out other influences (and noise) and isolate the ENSO signal, one can average DJF precipitation during El Niño or La Niña years.

Atmospheric variables vary from day to day, week to week, season to season, and so on. To describe these fluctuations—which collectively constitute climate variability—it is common to subtract the climatological mean value from a specific temporal value. This approach helps isolate and analyze the deviations that characterize climate variability.

$$Q' = Q - \bar{Q}$$

In this equation, Q is any variable with temporal variation and $\bar{Q}$ is its corresponding temporal average. Q′ defines the temporal perturbation of Q, also referred to as anomaly in literature. It is evident that the temporal average of Q′ is zero.

As we will discuss later, temporal disturbances can in turn be decomposed into certain temporal ranges to facilitate their description and study (low frequency, intraseasonal, daily, etc.). In this way the general circulation of the atmosphere can be described through the contribution of both the mean field and the different types of temporal perturbations.

The global or large-scale circulation also exhibits distinct spatial variations, primarily in the latitudinal and vertical directions, and to a lesser extent in the longitudinal direction. As a result, longitudinal (zonal) averages and vertical averages are frequently used in studies of atmospheric general circulation. A "zonal asymmetry" can then be defined as the deviation of the variable with respect to its latitudinal or zonal average:

$$Q^* = Q - [Q]$$

Stationary waves are a particular case of zonal asymmetry defined from the temporal mean of the variable:

$$\bar{Q}^* = \bar{Q} - [\bar{Q}]$$

In addition to averages, other statistical measures are useful for characterizing perturbations. For example, the variance ($\overline{Q'^2}$ or $[Q^{*2}]$), or its square root, the standard deviation, are commonly used to quantify the characteristic magnitude of fluctuations.

Covariance is used to describe the temporal relationship between two different variables. When one of the variables is a wind component, the covariance provides a measure of the transport or flux of the given variable Q by the wind perturbation.

2.1.2 Some Balance Equations

Given that this book is not aimed at graduates or advanced students in climate sciences, it is convenient to introduce some basic concepts that will be referenced throughout the following chapters.

The spatial description of the atmospheric and oceanic variables is carried out considering a coordinate system, in which the x axis is parallel to the circles of latitude and positive towards the east, and the y axis is parallel to the meridians, with zero value at the equator and positive values towards the north. Pressure may be used as the vertical coordinate, with values increasing downward toward the surface. Such a system of coordinates is known as the (x, y, p) system and is widely used in the atmospheric sciences community. Instead of comparing pressure at fixed altitudes, we compare altitudes or "geopotential heights" at fixed pressure levels.

In this system, the wind is represented through three components: u component or zonal component along the x axis and with positive values towards the east, v component or meridional component along the y axis with positive values towards the north, and ω component or vertical component along the vertical axis and with positive values indicating downward motion (i.e., descent toward higher pressure levels).

Under certain conditions, the behavior of the atmosphere approaches certain balanced states that are useful for describing and studying it.

Hydrostatic Balance

This balance is based on the vertical equilibrium between the upward pressure gradient force (also called pressure or buoyancy force) and the downward force of gravity. The full 3-dimensional pressure gradient force arises from the large spatial pressure variations on both vertical and horizontal scales in the middle and high latitudes. On large horizontal scales, that is, several hundred kilometers, the vertical velocity and its corresponding acceleration are much weaker than the horizontal ones, implying that the downward push of gravity is almost perfectly balanced by the upward buoyancy force due to the vertical decrease in pressure.

Using the hydrostatic balance and the equation of state of dry air, one can show that the thickness of an atmospheric layer between two pressure surfaces is proportional to the vertically-averaged temperature within that layer (warmer air is less dense and therefore takes up more volume). Since the atmosphere is warmer at the

Equator than at the poles, particularly in the lower to mid troposphere, pressure surfaces tilt downward towards the poles.

Geostrophic Balance

The horizontal movement that results from the balance between the horizontal pressure force and the Coriolis force is called the geostrophic wind. The fact that planet Earth is essentially spherical and rotates gives rise to an apparent (or inertial) force called the Coriolis force. This force is not real but results from the need to describe motion within a coordinate frame that rotates with the planet. The Coriolis force acts perpendicular to the direction of motion, deflecting flow to the left in the Southern Hemisphere and to the right in the Northern Hemisphere. In regions with strong horizontal pressure gradients, air parcels initially accelerate from high pressures to low pressures due to the pressure gradient force. When this force is balanced by the Coriolis force, the resulting motion or geostrophic wind is parallel to the isobars (lines of equal pressure), leaving the low pressures to the right of the flow in the Southern Hemisphere and to the left in the Northern Hemisphere. The intensity of this geostrophic wind is proportional to the horizontal pressure gradient: greater pressure differences result in stronger geostrophic winds. The geostrophic balance is a good approximation to the real wind outside the tropics and away from the surface (where friction results in a component towards low pressure). Applying this balance helps to understand and interpret the climatology and variability of the extratropical circulation. For instance, winds in midlatitudes are generally westerly (eastward) because pressures are higher in the subtropics than at the poles as explained above.

Thermal Wind Balance

The difference between the geostrophic wind at two pressure levels is known as the thermal wind (though it is not a real wind) and is proportional to the horizontal temperature gradient of in the layer. Since each successive atmospheric layer is warmer at the Equator than at the poles, the tilt of the pressure surfaces increases with height and so do the associated geostrophic westerly winds; thus, the thermal wind in midlatitudes is also westerly. A vertical shear of the geostrophic wind must be accompanied by a horizontal temperature gradient across the flow, with colder air to the right of the wind in the Southern Hemisphere (and to the left in the Northern Hemisphere). This dynamical constraint between the temperature and wind fields in the extratropics is called the thermal wind balance and provides a useful framework to interpret climate variability in extratropical and subpolar latitudes.

2.2 Relevant Characteristics of the Global Climate

2.2.1 Heat Balance in the Atmosphere

The "climate system" or "Earth system" encompasses the atmosphere, the oceans, the cryosphere and the land. Although the Earth's atmosphere extends hundreds of kilometers above the surface, the weather and climate that directly affect the surface occur in the lowest 10–15 km, a layer known as the troposphere, which contains about 80% of the atmosphere's mass. The atmosphere interacts with the underlying surface—which includes ocean, snow/ice and land areas, with varying vegetation, albedo (reflection of solar radiation) and surface roughness—through exchanges of momentum, water, heat, and gases such as carbon dioxide and methane. The surface also contributes aerosols or particles to the atmosphere through processes such as erosion, volcanic eruptions, and human activities.

The primary energy source driving the climate system is incoming solar radiation. This shortwave radiation is partially absorbed in the stratosphere (the layer above the troposphere) due to its high ozone content, which effectively filters out much of the Sun's ultraviolet energy. The remaining solar radiation passes through the troposphere largely unabsorbed. While some of this radiation is reflected by clouds and some is scattered by the atmosphere, about half of the incident radiation at the top of the atmosphere reaches the Earth's surface, which then heats and emits infrared radiation upward. A small fraction of this radiation escapes directly to space but most of it is absorbed by greenhouse gases such as carbon dioxide (CO_2) and water vapor, warming the troposphere. The combined longwave radiation emitted to space by the surface and atmosphere (or outgoing longwave radiation, OLR) offsets the incoming solar energy, maintaining a stable equilibrium temperature on Earth.

For purposes of intuitive understanding of how the earth cools by the emission of longwave radiation to space, it is convenient to think of the atmosphere as several horizontal layers. Each of these layers absorbs both incoming downward and upward longwave radiation and each layer also emits longwave radiation in both directions. The absorptivity and emissivity of the layer depends on the quantity of greenhouse gases in that layer. The amount of OLR escaping to space is thus a weighted sum of the contribution from each of the layers. If all layers have an equal amount of absorbing gases, the lowest layers emit more longwave radiation than the upper ones because emission depends on the fourth power of temperature and temperature decreases with height. Yet the contribution from the upper layers to the OLR is greater than that of the lower layers, because radiation from lower layers is more likely to be absorbed by layers above.

Two examples serve to illustrate extremes in OLR values. The first occurs over a dry atmosphere, typically associated with a hot, arid surface, such as a desert. In this case little radiation emitted by the surface is absorbed by this clear atmosphere, so OLR is large, as the top of the atmosphere effectively "sees" the hot surface. The second example is a region covered by convective clouds, which behave approximately like black bodies. In this case the cloudy atmosphere is nearly opaque to longwave radiation, so virtually no radiation emitted by the surface reaches space. All the outgoing radiation is emitted by the cold cloud tops, resulting in a reduced OLR.

Although globally the net incoming solar radiation at the top of the atmosphere must be balanced by OLR, locally the radiative balance at the top of the atmosphere varies significantly between the Equator and the Poles. While OLR is relatively uniform across latitudes, incoming solar radiation decreases sharply towards the poles. The resulting radiative imbalance must be compensated by a poleward transport of heat. This transport is achieved by the general circulation, which carry warm fluids poleward and cold fluids equatorward.

In considering this large-scale redistribution of energy, one must also include the energy exchanges that occur between the atmosphere and the surface. Most of the energy reaching the oceanic surface is used to evaporate water. Water vapor in the atmosphere acts as a heat reservoir, releasing its "latent" heat when it condenses as air rises and cools.

Estimates of the vertical distribution of diabatic heating (which includes contributions from radiative heating and cooling, release of latent heat associated with phase transitions, heating by conduction or sensible heating, and heating by frictional dissipation) for both DJF and JJA are shown in Fig. 2.1.

In the tropical troposphere of the summer hemisphere, strong heating is evident, primarily driven by the release of latent heat from water vapor condensation, with a pronounced seasonal migration and stronger values in the Northern Hemisphere. In extratropical latitudes, heating occurs primarily near the surface, with an additional weak heating maximum at mid-levels concentrated in a narrow midlatitude band, associated with synoptic weather systems. Outside these regions, the atmosphere predominantly loses energy through radiative cooling, meaning it absorbs less radiation than it emits. It is important to note that these estimates are derived indirectly from reanalysis or numerical models rather than direct measurements, so their quantitative values should be interpreted with some caution.

2.2.2 *Zonal Mean Circulation*

The spatial distribution of atmospheric heating plays a pivotal role in shaping temperature distribution. Throughout the year, the highest zonal mean temperatures are observed in the lower equatorial troposphere. The temperature distribution exhibits notable seasonal north-south migrations, which respond to variations in incoming solar radiation. In the tropics, zonal mean temperatures display relatively small meridional variations but exhibit strong vertical gradients. In contrast, horizontal temperature variations become much more pronounced in the mid-latitudes. In the Southern Hemisphere, they persist year-round, and in the Northern Hemisphere, they are most evident during winter.

The mean atmospheric circulation responds to these temperature contrasts in a manner consistent with the behavior of a rotating fluid on a spheroid (see Sect. 2.1).

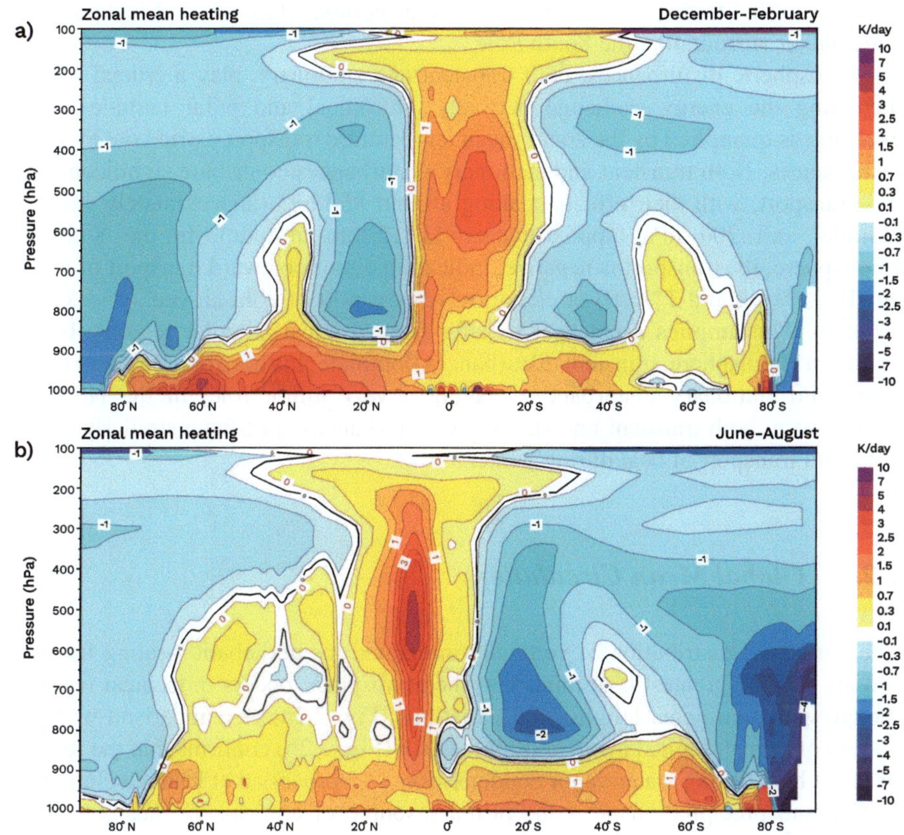

Fig. 2.1 Zonal-mean heating for (**a**) December–January-February, and (**b**) June–July-August. Units in °K/day (Adapted from ERA-40 Atlas, Kållberg et al. (2007))

In the meridional-vertical plane, the zonal mean circulation is dominated by the Hadley cell, a large-scale overturning circulation characterized by ascending motion in the Equator, poleward flow at upper levels and descending motion in the subtropics. The Hadley cell is most intense in the winter hemisphere, where the temperature gradient (and the resulting pressure gradient) between the equator and the poles is strongest. The Earth's rotation exerts a significant influence on this circulation, causing the upper poleward branch of the Hadley cell to deflect eastward in the subtropics. This deflection, combined with the upward intensification of westerly winds driven by horizontal pressure gradients, results in the formation of a maximum in zonal wind in the upper troposphere, known as the "subtropical jet" (recall the thermal wind balance). This jet is typically located around 30° latitude and is more intense in the winter hemisphere. At higher latitudes, another maximum in zonal wind, referred to as the "subpolar jet" or "eddy-driven jet," emerges due to interactions with transient atmospheric disturbances. While the eddy-driven jet persists in the troposphere year-round, during winter an additional westerly jet develops in the stratosphere as part of the polar vortex, a large area of low pressure and cold air surrounding the South Pole.

Atmospheric disturbances, both transient and stationary, play a crucial role in mediating the energy exchange between subtropical and polar latitudes. This exchange is quantified by the zonal-mean temperature transport carried out by these disturbances. Both transient and stationary disturbances contribute significantly to this transport, with their effects peaking around 850 hPa (approximately 1500 m altitude) and 200 hPa (approximately 11–12 km altitude). In the Southern Hemisphere, the transport is negative, indicating either poleward transport of warm air (i.e., $v^* < 0$ and $T^* > 0$) or equatorward transport of cold air (i.e., $v^* > 0$ and $T^* < 0$). This implies a net heat transfer from tropical to polar regions. In the Southern Hemisphere, transient disturbances dominate this transport throughout the year, with stationary disturbances playing a very minor role. In the Northern Hemisphere, both transient and stationary disturbances contribute similarly to the poleward transport of heat during winter.

2.2.3 *Global Mean Circulation*

The horizontal distribution of vertically integrated mean diabatic heating for DJF reveals that the greatest heat input into the atmosphere occurs in tropical regions, though with significant longitudinal variations (Fig. 2.2a). A comparison with the pattern of climatological mean rainfall for DJF (Fig. 2.2b) confirms that the release of latent heat during water vapor condensation — a process that leads to cloud formation and precipitation — is the primary source of atmospheric heating in the tropics. In these regions, intense convergence of mass and moisture in the lower troposphere drives upward air motions, leading to cooling and condensation (Fig. 2.2c). In the equatorial Pacific and Atlantic Oceans, narrow bands of moisture convergence and rainfall, known as the Intertropical Convergence Zones (ITCZs), mark the ascending branches of the Hadley cell. Over continents, rainfall patterns are influenced by monsoon systems, which are driven by land-sea thermal contrasts. The rainfall bands associated with the South Pacific Convergence Zone (SPCZ) and the South Atlantic Convergence Zone (SACZ) are also evident over the southwestern Pacific and southwestern Atlantic, respectively. During JJA, the regions of maximum diabatic heating, rainfall, and moisture convergence shift northward, following the seasonal migration of solar radiation (not shown).

The horizontal distribution of upper-tropospheric mean winds also exhibits significant seasonal and longitudinal variations. In the Southern Hemisphere, the JJA subtropical jet is most pronounced between Australia and the west-central South Pacific. The longitudinal differences in jet intensity reflect variations in the strength of the Hadley cell, in turn closely linked to differences in diabatic heating. In both seasons, the subpolar jet in the mid-to-high latitudes is prominent, with its maximum intensity over the southern Indian Ocean. This jet plays a crucial role in

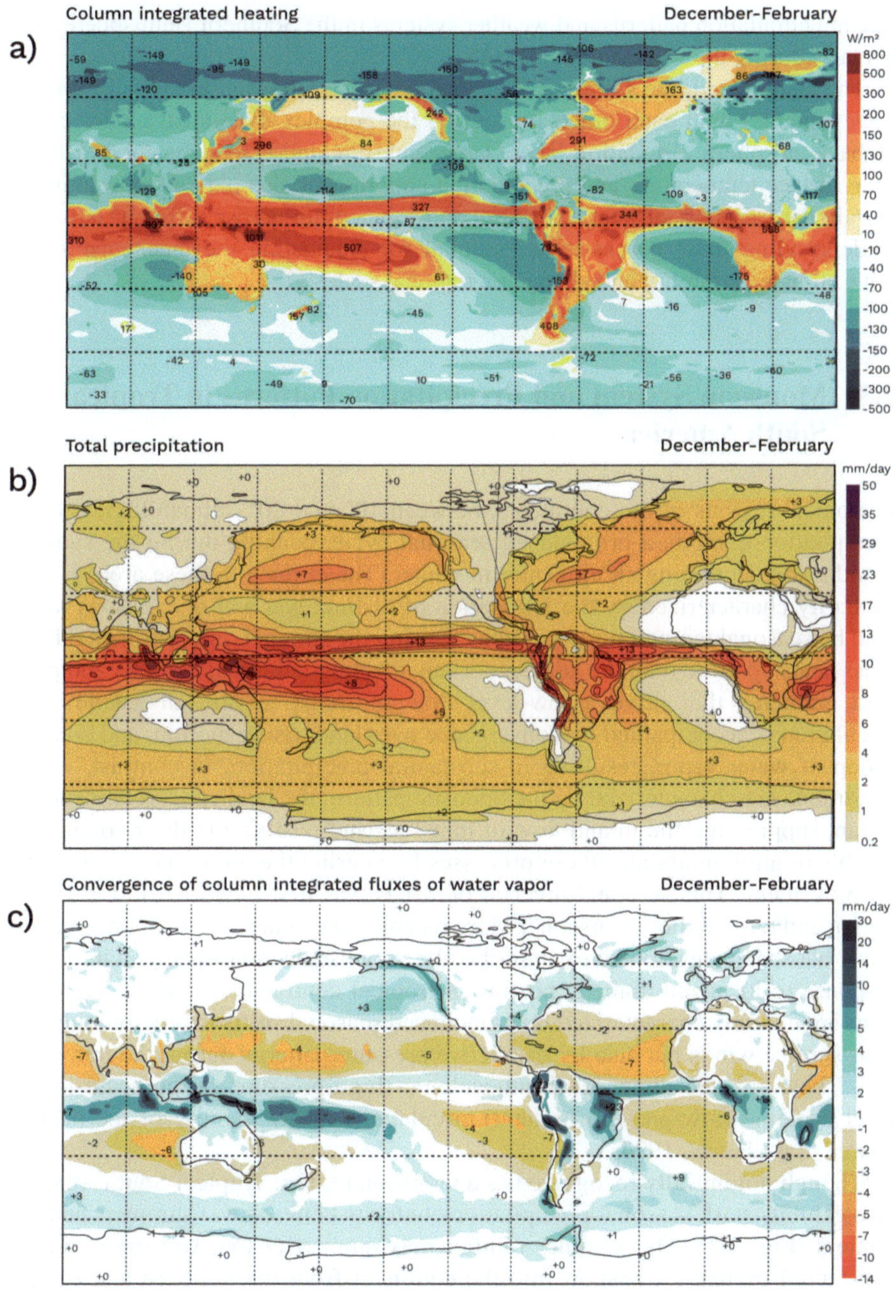

Fig. 2.2 Means of (**a**) column integrated heating (units in W/m²), (**b**) total precipitation (units in mm/day) and (**c**) convergence of column integrated fluxes of water vapor (units in mm/day) (Adapted from ERA-40 Atlas, Kållberg et al. (2007))

shaping circulation patterns and weather systems in the Southern Hemisphere, particularly during the winter months (JJA).

The Hadley cell, the subtropical and subpolar jets, the stationary and transient waves interact with each other, illustrating the complex dynamical nature of the global atmospheric circulation. These large-scale circulation features not only influence surface temperature and precipitation patterns but also play a key role in the meridional transport of heat, moisture, and momentum across different latitudes. Understanding these processes is essential for elucidating the mechanisms that drive climate variability and change on regional scales.

2.3 Monsoon System and Precipitation Patterns in Tropical South America

In tropical regions of the continents, between 10° and 20°, climatological mean precipitation exhibits a pronounced seasonal cycle, peaking during the summer months and decreasing significantly in winter. The climate in these areas is then primarily characterized by the alternation of distinct wet and dry seasons rather than by seasonal temperature variations. During the summer, the continental landmass heats up more than the adjacent ocean, creating a thermal contrast and pressure gradients that drive a low-level flow of moist marine air inland. This circulation is accompanied by convergence over land, rising motion, and condensation of water vapor, resulting in substantial precipitation. In contrast, winter conditions are characterized by the opposite pattern: the land cools relative to the ocean, suppressing the inland flow of moist air and resulting in drier conditions. Earth's rotation modulates these processes by altering the direction of horizontal air movement. This seasonal reversal of winds was first identified in the Indian region and termed the "monsoon". More recently, the concept of a "monsoon system" has been developed to describe the large-scale atmospheric circulation response to localized diabatic heating in tropical continental regions. Such systems are not only characterized by pronounced seasonal shifts in precipitation and near-surface winds but also by distinctive pressure patterns during the rainy season. These include upper-level anticyclonic (high-pressure system) and lower-level cyclonic (low-pressure system) circulations in the vicinity of the heating source (e.g., James, 1994).

Though not initially recognized as a monsoon region, Tropical South America also exhibits stark seasonal precipitation contrasts between wet summers and dry winters (Fig. 2.3). The summer rainfall maximum over the Amazon is due to substantial moisture transport into the continent from the northern and eastern coasts. This moisture transport is deflected southward by the presence of the Andes Cordillera, feeding into a low-level continental anticlockwise circulation gyre (Fig. 2.3a). During winter, when rainfall over central South America diminishes significantly, the continental circulation gyre persists, although weaker (Fig. 2.3b).

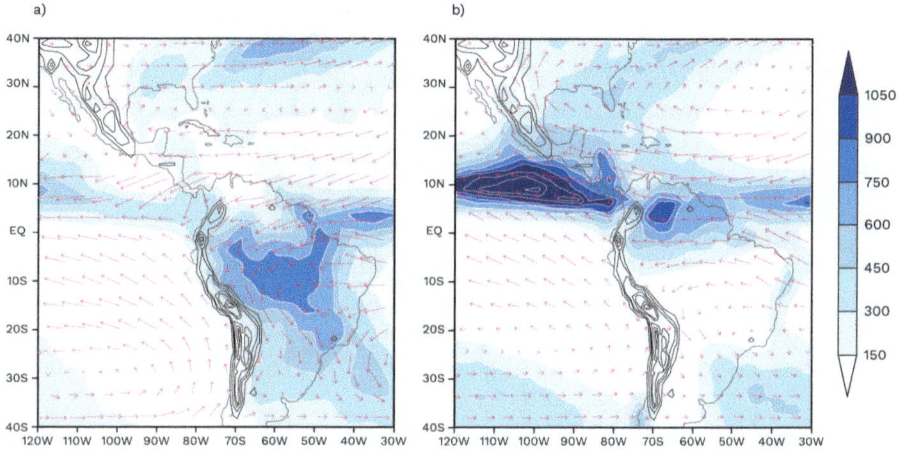

Fig. 2.3 Mean accumulated precipitation (shaded) (From Xie and Arkin (1997)) and vertically averaged climatological mean moisture fluxes (arrows) (from NCEP–NCAR reanalysis) for (**a**) DJF and (**b**) June–August (JJA). Contour interval is 150 mm. Vector units are kg (m s)$^{-1}$. Orography higher than 1000 m is contoured in black (Adapted from Vera et al. (2006a))

The absence of a clear reversal in near-surface winds between the rainy and dry seasons led the scientific community to initially conclude that South America did not have a monsoon system. In the late 1990s, studies such as that of Zhou and Lau (1998) demonstrated that the persistent continental anticyclonic gyre throughout the year results from mechanical forcing by the Andes. However, when the annual mean is subtracted from the seasonal mean for winter and summer, a distinct seasonal reversal of low-level winds becomes evident, driven by the varying thermal contrast between the continent and the ocean, as expected in a monsoonal system (Zhou & Lau, 1998).

The South American monsoon system is uniquely shaped by the presence of the north-south oriented Andes Mountains (Vera et al., 2006a). Furthermore, the extensive tropical landmass of South America distinguishes it from the Indian monsoon system, since much of the Indian subcontinent lies within extratropical latitudes. In South America, moisture transported from the Atlantic Ocean must cross a vast continental area within the tropics. Consequently, the regional climate dynamics are influenced not only by sea-atmosphere interactions in the tropical Atlantic but also by land-atmosphere interactions, particularly with tropical vegetation, such as that of the Amazon rainforest.

A closer examination of summer rainfall distribution (Fig. 2.3a) reveals that the maximum over the Amazon at around 10°S is connected to the SACZ, the band of relatively high rainfall extending southeastward. As discussed in later chapters, the SACZ plays a critical role in shaping climatic conditions across the eastern portion of the continent, including southeastern Brazil and the SESA region further south.

Another key regional feature is the South American low-level Jet (SALLJ), which refers to a narrow intense current of moist air flowing southward along the eastern slopes of the Andes over Bolivia at altitudes between 1000 and 1500 meters,

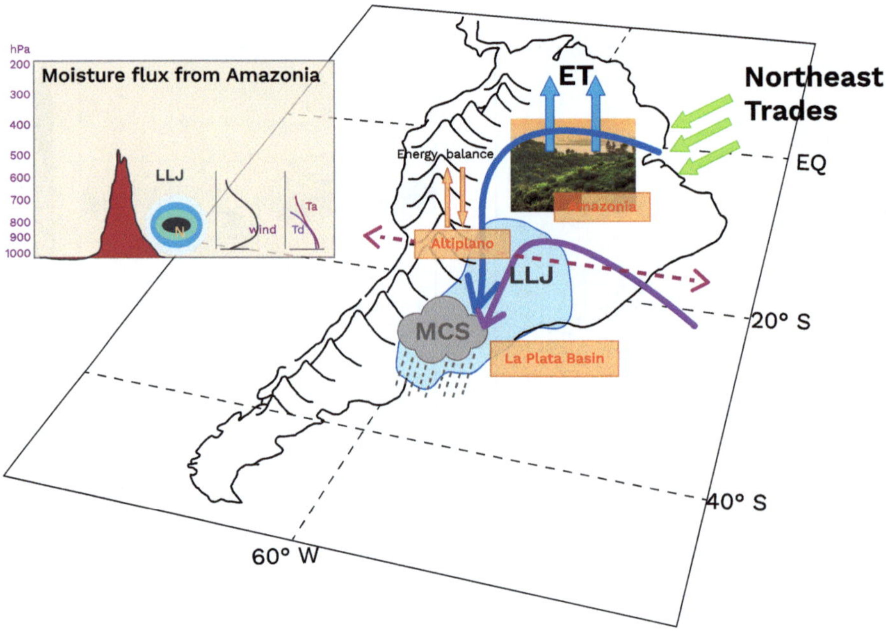

Fig. 2.4 Schematic diagram of elements relevant to poleward moisture transport over South America. Blue and purple curved arrows depict the moisture transport into the continent from the tropical and South Atlantic Ocean, respectively. ET means evapotranspiration and MCS means mesoscale convective systems. The inset represents a vertical cross section of the northerly flow along the dashed line displayed in the diagram, including wind, temperature and dew-point temperature profiles representative of the LLJ core (Adapted from Vera et al. (2006b))

where the moisture flux shown in Fig. 2.3 is concentrated. Subtropical rainfall is enhanced by this channeling of the moisture flow from the Amazon to the LPB, a process sustained by the presence of the Andes (Fig. 2.4). Although strong SALLJ events occur on only 17% of summer days, they contribute to over 50% of the total precipitation in subtropical LPB (Salio et al., 2002).

The SALLJ's three-dimensional structure (Fig. 2.4) was first comprehensively described during the SALLJEX field experiment in the summer of 2002–2003 (Vera et al., 2006b). This campaign utilized aircraft observations, complemented by radiosondes and pilot balloon data, to confirm the existence of the SALLJ, which had previously been inferred from theoretical calculations, limited local data, and numerical model simulations. At higher atmospheric levels, the summer circulation over South America is characterized by a prominent high-pressure center or anticyclonic circulation centered at about 20°S, known as the Bolivian High. This feature is the high-altitude response to the diabatic heating generated by rainfall over tropical South America.

As will be discussed in subsequent sections, the various components of the South American Monsoon System, including the subtropical regions, exhibit significant variability across multiple timescales, ranging from hours and days to years.

2.4 Seasonal and Diurnal Cycles in South America

Rainfall in South America exhibits distinct seasonal variations, with characteristics that vary depending on location (Vera et al., 2006a). In the Amazon basin, the rainy season begins around September, preceded by an intensification of the northerly moisture flow along the continent's northern coast. Between September and October, rains typically commence in central and southeastern Brazil, while from November to February the SACZ develops, accompanied by heavy rains over the Bolivian Altiplano. From March to April, rainfall in central and southeastern Brazil decreases, signaling the start of the dry season, which persists until the following spring. The northern LPB, located within the tropical zone, experiences well-defined wet and dry seasons. In contrast, the basin southern portion, located in the subtropical region of SESA, receives substantial rainfall year-round, with slightly higher amounts in autumn and spring.

Precipitation in the tropical regions of South America, as well as in the subtropical regions during the warm season, is predominantly convective, characterized by showers generated by cumulus-type clouds. These clouds develop due to the vigorous ascent of warm, humid air masses (WMO, 2017). Convection exhibits a strong diurnal cycle, closely linked to temperature variations and driven by thermodynamic processes. Additionally, the timing of the rainfall diurnal cycle in these regions varies depending on geographic location and the influence of topographic features such as mountain ranges or coastlines (Vera et al., 2006a).

By using satellite data and numerical models, which complement the limited sub-daily rainfall observation records, researchers have been able to characterize the diurnal cycle of summer rainfall across the continent. During the afternoon and evening, convective clouds are prevalent in the Bolivian Altiplano, along the northeastern coast, and over the central Amazon basin. Convection also develops in the afternoon across central-southeastern Brazil and the SACZ, including the northern part of the LPB. In contrast, nocturnal and early morning convection is prominent along the eastern slopes of the subtropical Andes, extending eastward into the subtropical section of the LPB (Vera et al., 2006a).

2.5 Mesoscale and Synoptic Disturbances in Southern South America

The occurrence of precipitation in the LPB is closely linked to the development and propagation of convective storms, often triggered by local-scale or mesoscale disturbances. These disturbances, in turn, are driven or modulated by transient atmospheric systems that evolve over periods of a few days (synoptic scale). Such weather systems enhance the convergence of near-surface moisture fluxes, leading to the ascent and subsequent condensation of water vapor (Vera et al., 2006a).

In SESA, mesoscale convective systems account for approximately 60% of the total rainfall (e.g., Durkee et al., 2009). As previously noted, these storms typically develop in the evening and reach their peak intensity at night. Compared to other regions, the convective systems in SESA are notably larger and longer lasting, with an average duration of around 12 h (Salio et al., 2007). SALLJ activity plays a critical role in the formation of these subtropical storms. During the summer months, the SALLJ exhibits significant diurnal variability, with intensity peaking in the late afternoon and evening. This diurnal behavior creates favorable conditions for nocturnal convection in SESA (e.g., Nicolini et al., 2004, Zipser et al., 2004).

Cyclones and anticyclones are among the most relevant synoptic-scale atmospheric disturbances in mid-latitudes. In the Southern Hemisphere, these systems develop and propagate in the extratropics from west to east year-round, with their influence on the southern portion of South America being more pronounced during winter (Vera et al., 2006a). The presence of the Andes Mountains forces the low-level steering flow to bend equatorward and enhances the northward reach of these synoptic systems at lower levels. When cyclones interact with the mountain range, they intensify and extend the SALLJ into the LPB, promoting favorable conditions for precipitation in that region. Conversely, anticyclones moving across the Andes can drive equatorward intrusions of dry and cold air, occasionally leading to frost events, even in tropical regions.

References

Durkee, J. D., Mote, T. L., & Marshall Shepherd, J. (2009). The contribution of mesoscale convective complexes to rainfall across subtropical South America. *Journal of Climate, 22*, 4590–4605. https://doi.org/10.1175/2009JCLI2858.1

Hersbach, H., Bell, B., Berrisford, P., Hirahara, S., et al. (2020). The ERA5 global reanalysis. *Quarterly Journal of the Royal Meteorological Society, 146*(730), 1999–2049. https://doi.org/10.1002/qj.3803

James, I. (1994). *Introduction to circulating atmosphere*. Cambridge University Press.

Kållberg, P., Simmons, A., Uppala, S., & Fuentes, M. (2007). ERA-40 project report series, 17. *The ERA-40 Archive*, 31 pp, https://www.ecmwf.int/node/10595

Kalnay, E., Kanamitsu, M., Kistler, R., et al. (1996). The NCEP/NCAR 40-year reanalysis project. *Bulletin of the American Meteorological Society, 77*(3), 437–471. https://doi.org/10.1175/1520-0477

Nicolini, M., Salio, P., Ulke, G., et al. (2004). South American low-level jet diurnal cycle and three-dimensional structure. *CLIVAR Exchanges, 9*(1), 6–9. International CLIVAR Project Office, Southampton, United Kingdom.

Salio, P., Nicolini, M., & Saulo, A. C. (2002). Chaco low-level jet events characterization during the austral summer season. *Journal of Geophysical Research, 107*(D24), 4816. https://doi.org/10.1029/2001JD001315

Salio, P., Nicolini, M., & Zipser, E. J. (2007). Mesoscale convective systems over southeastern South America and their relationship with the south American low-level jet. *Monthly Weather Review, 135*, 1290–1309. https://doi.org/10.1175/MWR3305.1

Vera, C., Higgins, W., Amador, J., et al. (2006a). Toward a unified view of the American monsoon systems. *Journal of Climate, 19*, 4977–5000. https://doi.org/10.1175/JCLI3896.1

Vera, C., Baez, J., Douglas, M., et al. (2006b). The South American low-level jet experiment (SALLJEX). *Bulletin of the American Meteorological Society, 87*, 63–77. https://doi.org/10.1175/BAMS-87-1-63

WMO. (2017). *International cloud atlas: Manual on the observation of clouds and other meteors* (WMO-No. 407). World Meteorological Organization. https://library.wmo.int/index.php?lvl=notice_display&id=5357

WMO. (2019). *Guide to the WMO integrated global observing system* (WMO-No. 1165). World Meteorological Organization. https://library.wmo.int/index.php?lvl=notice_display&id=20026

Xie, P., & Arkin, P. A. (1997). Global precipitation: A 17-year monthly analysis based on gauge observations, satellite estimates, and numerical model outputs. *Bulletin of the American Meteorological Society (BAMS), 78*, 2539–2558. https://doi.org/10.1175/1520-0477(1997)078<2539:GPAYMA>2.0.CO;2

Zhou J, WK-M Lau WK-M (1998) Does a monsoon climate exist over South America? Journal of Climate 11:1020–1040. https://doi.org/10.1175/1520-0442(1998)011%3C1020:DAMCEO%3E2.0.CO;2.

Zipser, E. P., Salio, P., & Nicolini, M. (2004). Mesoscale convective systems activity during SALLJEX and the relationship with SALLJ events. *CLIVAR Exchanges, 9*(1), 14–18. International CLIVAR Project Office.

Chapter 3
Intraseasonal Variability

3.1 Basic Concepts of Intraseasonal Variability

Decision-making in many climate-sensitive socioeconomic sectors—such as disaster risk management, logistics planning and agriculture— relies on anticipating climate conditions several weeks in advance. Climate fluctuations occurring on timescales between roughly 2 weeks and a season are broadly referred to as intraseasonal oscillations (ISOs), or, more generally, as intraseasonal variability (ISV). Definitions vary in literature, with some studies referring to periodicities between 20 and 100 days (e.g., Lau & Waliser, 2011), while others cite ranges of 8–70 days (e.g., Liu et al., 2022).

Regardless of its precise definition, ISV occurs on timescales that lie between those associated with synoptic-scale phenomena (such as cyclones, anticyclones, and cold fronts, which have periodicities of up to 7–8 days) and longer-term climate variability (which operates across seasons and beyond). As a result, characterizing ISV requires application of mathematical filters to time series of atmospheric variables to separate their signals from shorter and longer period climate oscillations (e.g., James, 1994). This process reveals that ISV accounts for approximately 20 to 30% of warm-season precipitation variability in tropical and subtropical regions across all continents.

Research on ISV has been driven largely by the need to predict its behavior. Predicting climate conditions on timescales of a few weeks—often referred to as "medium-term" or "extended-range" prediction—has emerged as a significant challenge for the scientific community in recent years, within the framework of "subseasonal prediction." Since the 1970s, short-term predictions (up to approximately 7 days) have been made using numerical models that solve the fundamental equations governing the climate system, based on the instantaneous state of the atmosphere (the initial conditions described in 2.1). Efforts at seasonal prediction, on the other hand, began to accelerate in the 1990s. These predictions are based on the

C. S. Vera et al., *Climate Variability and Change in Southern South America*, SpringerBriefs in Environmental Science,
https://doi.org/10.1007/978-3-032-07197-2_3

understanding that atmospheric variations on these time scales are influenced by slowly varying components of the climate system rather than by initial conditions, such as the state of the oceans and land surface (known as boundary conditions), and even the stratosphere. ISV behavior is influenced by both the initial atmospheric conditions and these boundary conditions. To advance predictions on intermediate or intraseasonal timescales, it has been necessary to deepen our understanding of ISV behavior, its sources of predictability, and the associated dynamical processes at both global and regional scales.

Much of ISV activity can be explained by tropical climate dynamics—including equatorial waves, boundary layer processes, and convective coupling—as well as cloud–radiative feedbacks and air-sea interactions. In particular, the Madden-Julian oscillation (MJO; Madden & Julian, 1972), a coupled convective-dynamic tropical wave, is the dominant tropical pattern of ISV (see next section). Tropical ISOs may also be affected by marine biology and atmospheric chemistry, among others (Lau & Waliser, 2011).

The influence of tropical ISV extends beyond the tropics to extratropical and subpolar regions, through various teleconnection pathways (discussed in Sect. 3.3). ISV can also originate in extratropical latitudes, because of fluctuations in the mean westerly flow. For example, when the westerly jet weakens, a large-scale quasi-stationary high-pressure anomaly that 'blocks' the normal westerly flow may occur, sometimes persisting for days or even weeks and significantly altering regional climate conditions (Sect. 3.4). The remainder of the chapter will focus on the main characteristics of ISV over South America and its impact on climate variability in the LPB. The predictive skill and remaining challenges of subseasonal forecasting are addressed in Chap. 6.

3.2 The Madden-Julian Oscillation

The MJO is the most significant oscillation on intraseasonal timescales and is characterized by a periodicity of approximately 30–60 days (Madden & Julian, 1972). Originating in the tropics, the MJO manifests as variations in a wide range of atmospheric variables, including winds at lower and upper levels, rainfall, cloudiness, and sea-level pressure. It is also detectable in oceanic variables such as SST, surface evaporation and chlorophyll concentration (Lau & Waliser, 2011 and references therein). The upper-level signal of the MJO propagates outside the tropics, influencing many regions in mid- and subpolar latitudes.

The MJO is a large-scale wave-like disturbance that propagates eastward along the equator. It features an anomalous rising branch accompanied by a descending branch to its east, with anomalous upper-level westerlies and lower-level easterlies in between (Fig. 3.1). Due to its broad zonal extent, when the MJO is active it essentially divides the tropics into two contrasting regions, one experiencing enhanced cloudiness and precipitation and the other suppressed convection. For instance, one may have enhanced convection over Southeast Asia and nearby archipelago regions

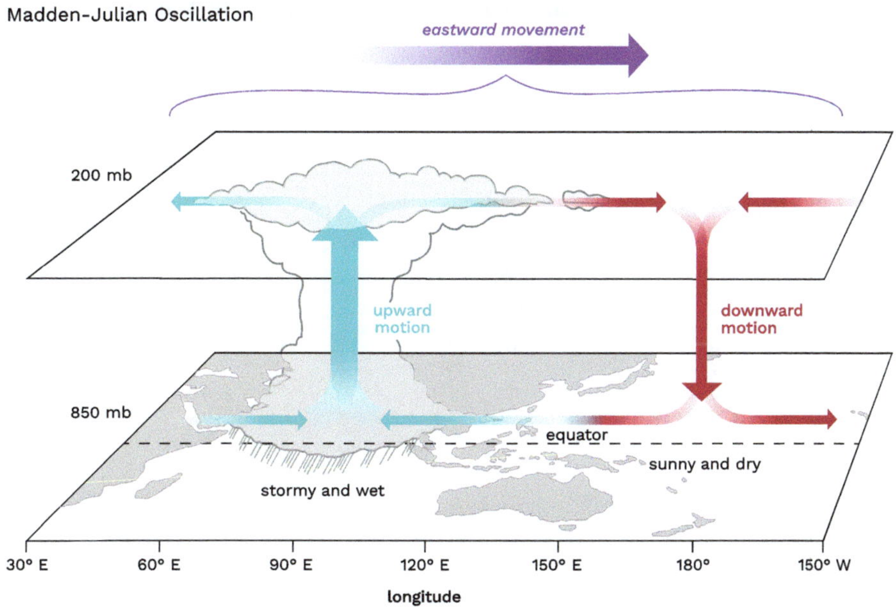

Fig. 3.1 The surface and upper-atmosphere structure of the MJO for a period when the enhanced convective phase (thunderstorm cloud) is centered across the Indian Ocean and the suppressed convective phase is centered over the west-central Pacific Ocean. Horizontal arrows pointing left represent wind departures from average that are easterly, and arrows pointing right represent wind departures from average that are westerly (Adapted from Climate.gov schematic, by Fiona Martin)

in the Indian and Pacific Oceans and weakened convection over northern and central South America, or vice-versa. The anomalies circle around the globe in about 30–60 days, moving faster in the Western Hemisphere than in the Eastern Hemisphere. To track this eastward progression, the MJO is commonly divided into eight phases (or stages), each representing a distinct longitudinal position of the convective center (e.g., Wheeler & Hendon, 2004).

The MJO is most prevalent during the late spring, summer, and early autumn of the Southern Hemisphere (approximately October to April) and tends to be more active during years with ENSO-neutral conditions (chapter 4). Even during peak phases, the MJO is highly irregular, with intervals of moderate to intense activity often followed by periods of weak or no activity.

The mechanisms driving the MJO can be broadly categorized into two groups: those related to external forcing (such as ISV activity in the Asian Australian monsoon or extratropical perturbations penetrating into the tropics) and those associated with internal variability (such as tropical convection and air-sea interaction) (Zhang, 2005). The MJO's influence on extratropical regions is indirect, mediated through changes in the local Hadley circulation (Chap. 2) or horizontal teleconnections that link the tropics to the extratropics, as discussed in the next section.

MJO activity is routinely monitored by global and regional climate centers, using indices that track its key features, such as the location and strength of anomalous

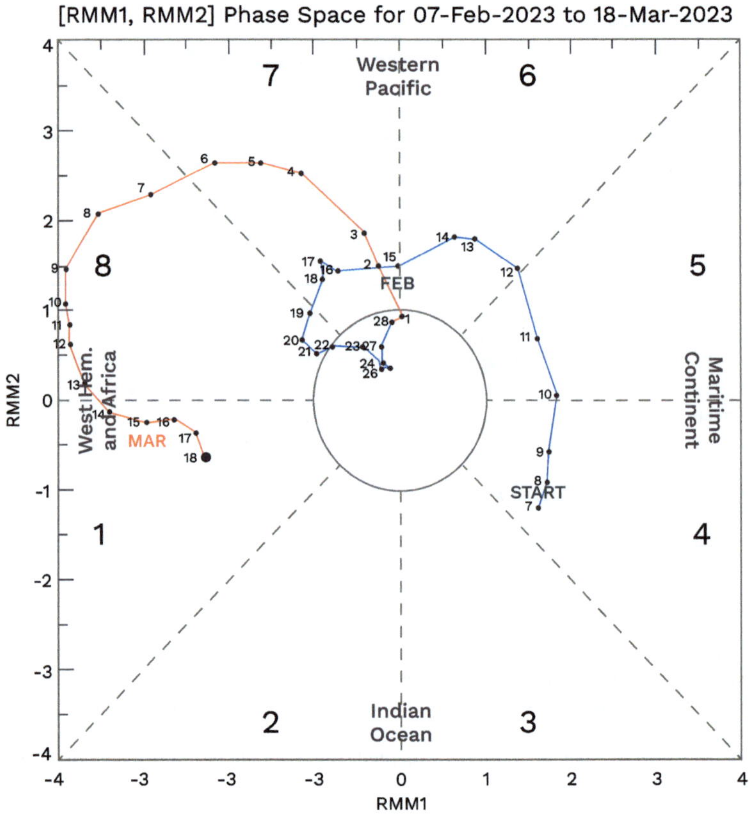

Fig. 3.2 Daily MJO index for 7 February 2023 to 18 March 2023 derived from Wheeler and Hendon (2004). The dot corresponding to each day indicates the intensity and phase of the MJO. A line connects the dots that is colored by month. The intensity is stronger the further the dot is from the center. The eight phases of the MJO are distinguished by the dotted lines and numbers (Adapted from CPC/NOAA MJO site)

convection. One of the most widely used indices is the multivariate MJO index, developed by Wheeler and Hendon (2004), which is based on a pair of empirical orthogonal functions (EOFs) of the combined fields of near-equatorially low-level zonal wind, upper-level zonal wind, and OLR data. The MJO index is typically visualized as a diagram depicting the eight phases of the oscillation (Fig. 3.2). Each point on the diagram corresponds to the index value for a given day, connected by a line that traces the oscillation's eastward progression. The eight phases are indicated by octants limited by dotted lines, colors represent different months, and the distance from the center indicates the intensity of MJO activity, with points near the center indicating weak or inactive periods.

MJO activity can be predicted to some extent, with global prediction centers producing both dynamical and statistical forecasts that provide skillful MJO outlooks up to 15–40 days in advance (e.g., Gottschalck et al., 2010).

3.3 Teleconnections Between the Tropics and Extratropics

Both tropical and extratropical regions serve as sources of ISV. Within the broad ISV spectrum, two distinct bands of activity can be identified based on their temporal characteristics (Liu et al., 2022 and references therein): variability with relatively short periods (~8–20/30 days) and relatively longer periods (~20/30–70 days). The latter is primarily associated with the MJO, while the former may originate either in the tropics, due to shorter-period equatorial oscillations and associated nonlinear processes (e.g., Raupp & Silva Dias, 2010), or in mid-latitudes from interactions between the westerly flow and perturbations (e.g., Liu et al., 2022).

The term "teleconnections" refers to statistical relationships between climate anomalies occurring in geographically distant regions. These relationships are often described through leading patterns of atmospheric circulation anomalies, represented by variables such as sea level pressure, geopotential height, streamfunction, and wind. Teleconnections typically manifest as quasi-stationary waves with relatively fixed spatial nodes and antinodes that oscillate on intraseasonal timescales. This alternation of anomalies with opposite signs is associated with the meridional propagation of Rossby waves.[1] The dynamics of the Rossby wave trains play a central role in connecting the tropics and extratropics, as well as linking polar and mid-latitude regions.

The MJO can force teleconnections in the Southern Hemisphere by inducing Rossby wave trains that typically originate in the west-central tropical Pacific Ocean or the tropical Indian Ocean, propagate poleward into subpolar regions and then turn equatorward toward South America and the Atlantic Ocean (e.g., Alvarez et al., 2016). For instance, during the MJO phase 7, the anomalous descending branch located over the tropical Indian Ocean triggers a wavetrain of anomalies that arches poleward before recurving toward South America, producing a large upper-level anticyclonic anomaly which in turn alters climate conditions in southern South America. (Fig. 3.3). A similar wave train but with reversed sign is observed in the opposite phase (3), when the anomalous ascent occurs over the tropical Indian Ocean. These wave trains, which span the entire South Pacific Ocean, are collectively referred to as the Pacific-South American (PSA) patterns (e.g., Mo & Higgins, 1998).

For ISOs with periods of 20–30 days or less, wave trains can originate in the mid-latitudes over the three ocean basins in the Southern Hemisphere and propagate equatorward over South America, Africa, and Australia (e.g., Liu et al., 2022). During summer, wave trains on these shorter intraseasonal timescales can also be excited by changes in convection linked to the SPCZ or to the SACZ, extending zonally along subtropical latitudes (e.g., Vera et al., 2018). Under favorable

---

[1] Rossby waves are atmospheric waves excited by internal atmospheric mechanisms or by external forcing such as orography or convection. They are associated with timescales ranging from days to weeks, seasons, and longer. On the longest timescales, Rossby wave trains can extend to planetary spatial scales and remain predominantly quasi-stationary.

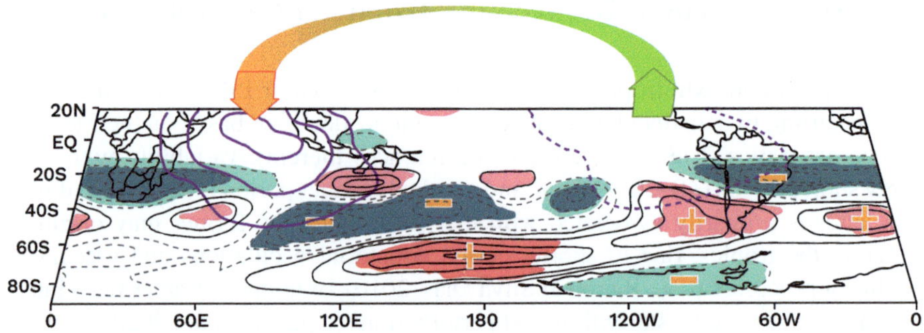

Fig. 3.3 Schematic of how the MJO influences JJA circulation during phase 7. The green arrow indicates the anomalous rising air that occurs over South America and the brown arrow indicates the anomalous descending air over southeast Asia and western Pacific. Solid contours indicate positive values, and dashed contours indicate negative values. The black contours along with pink and blue shading represent the higher and lower than normal pressure zones produced by the MJO, respectively. Violet contours at the tropics describe anomalous upper-level inflow (solid) and outflow (dashed) (Adapted from Alvarez et al. (2016))

atmospheric conditions, these wave trains may propagate toward the equator and even cross into the Northern Hemisphere, as observed in the eastern equatorial Pacific and equatorial Atlantic (Kiladis & Weickmann, 1992; Vera et al., 2018).

Not all teleconnections are associated with wave trains. Some instead involve alternating circulation anomalies between two distinct states or phases, typically anchored in specific geographical positions. A prominent example in the Southern Hemisphere is the SAM, which is characterized by a pressure anomaly over Antarctica (hence the term "annular"), flanked by two to four anomalies of opposite sign in the mid-latitudes, depending on the season (e.g., Thompson & Wallace, 2000). The positive phase of the SAM is associated with an intensification of the midlatitude westerly winds, while its negative phase corresponds to a weakening of these winds. The primary mechanism driving this SAM activity is internal atmospheric dynamics, particularly the interaction between the westerly flow and transient disturbances, which leads to fluctuations between stronger and weaker states of the extratropical jet.

The SAM exhibits variability across a wide range of timescales from synoptic and intraseasonal, to seasonal, interannual and longer (Lorenz & Hartmann, 2001). SAM variability on intraseasonal timescales alters the trajectories of cyclones and anticyclones in the Southern Hemisphere extratropics, which in turn affect the climate of southern South America (e.g., Carvalho et al., 2005). Moreover, the SAM has been shown to modulate MJO-induced teleconnections in the South Pacific, either by amplifying or dampening them (Flatau & Kim, 2013). Conversely, although tropical phenomena like the MJO may influence the SAM, conclusive evidence remains limited.

3.4 Blocking of the Westerly Flow in the Southern Hemisphere

A "blocking" condition in mid-latitude circulation is characterized by a persistent anticyclone or high-pressure region that remains stationary for several days (e.g., James, 1994). Once established, blocking disrupts the typical eastward progression of meteorological systems and their associated weather phenomena, such as cloudiness and rainfall. These systems are instead diverted around the blocking anticyclone, primarily on its poleward side.

Blocking events are less frequent in the Southern Hemisphere compared to the Northern Hemisphere. However, they are more likely to occur during the cold season in specific regions, such as the southwest Pacific, south of New Zealand, where the subpolar westerly jet weakens, and over the ocean areas surrounding southern South America, the southeastern Pacific and the southwestern Atlantic (Trenberth & Mo, 1985; Damião Mendes & Cavalcanti, 2014). The occurrence of one or more simultaneous blocking anticyclones is sometimes associated with the development of a wave-3 pattern (i.e., three waves or three pairs of opposite-sign anomalies in sea level pressure) across the Southern Hemisphere extratropics.

The occurrence and maintenance of blocking systems is linked to the interaction between the mean flow and transient disturbances. This interaction can weaken the westerly flow and favor the development of one or more simultaneous blocking anticyclones. These anticyclones, in turn, help sustain the weakened flow. Teleconnections between the tropics and extratropics can also alter the basic state of circulation, promoting the development of blocking anticyclones. This mechanism is particularly active in regions such as the Amudsen-Bellinghausen Sea and the Wedell Sea, surrounding the Antarctic Peninsula (Renwick & Revell, 1999).

SAM activity influences the frequency and duration of blocking events (Damião Mendes & Cavalcanti, 2014). During its positive phase, blocking events are more frequent and long-lived in the southwestern Pacific and southwestern Atlantic. In contrast, blocking activity in the southeastern Pacific is similar in both SAM phases, but events tend to persist longer during the negative phase, especially in El Niño years (Oliveira et al., 2014).

3.5 Main Patterns of Intraseasonal Variability in South America

ISV in southern South America is very pronounced year-round, with characteristics and mechanisms that vary seasonally (Vera et al., 2018 and references therein). OLR is commonly used as an indicator of regional ISV activity, serving as a reliable proxy for rainfall variations on intraseasonal scales in tropical and subtropical regions of the continent (e.g., Gonzalez et al., 2008). To describe the main regional patterns of ISV, EOF analysis is applied to filtered OLR anomalies in the intraseasonal band (Vera et al., 2018). The resulting leading pattern of ISV in southern South America is colloquially known as the Seasonal-Intraseasonal pattern (SIS), and its associated time series serves as an index of SIS activity (e.g., Alvarez et al., 2017). When the SIS index exceeds a magnitude of 1 standard deviation for a sustained period, a "SIS event" is identified. These events are classified as "short" if they last 3–5 days and "long" if they persist for longer. Using statistical methodologies such as linear regression, the activity of the SIS index can be linked to the behavior of climatic variable anomalies such as precipitation and temperature in southern South America (Vera et al., 2018). In this way, SIS events can be effectively used to characterize sequences of rainy days (i.e., wet periods) or dry spells with consecutive days without precipitation, as well as periods of above- or below-normal temperatures. SIS index can be defined either for the entire 10–90-day band or separately for both long (30–90 days) and short (10–30 days) timescales (Vera et al., 2018).

During the warm season (October to April), ISV in southern South America, including the LPB, is strongly influenced by the South American Monsoon, particularly by the activity of the SACZ. In this season, the SACZ region exhibits the strongest rainfall variability on intraseasonal timescales (Liu et al., 2022; Liebmann et al., 1999). The warm-season SIS pattern on both long (SIS_LP) and short (SIS_SP) timescales exhibits a dipole-like structure with an OLR anomaly center in the SACZ region and another of opposite sign in SESA (Vera et al., 2018 and references therein). Figure 3.4 illustrates the key characteristics associated with the two SIS phases. The positive phase of SIS is associated with (i) increased rainfall and negative near-surface air temperature anomalies in SESA, together with suppressed rainfall and positive temperature anomalies in the SACZ region; (ii) an extended and intensified SALLJ directed toward SESA; and (iii) anticyclonic pressure anomalies over southeastern Brazil and cyclonic anomalies near the southern tip of South America. The reverse occurs in the negative SIS phase.

ISV modulates the occurrence of extreme events in the LPB throughout the year (e.g., Hirata & Grimm, 2017). During positive SIS phases, extreme daily precipitation events are more frequent in SESA (e.g., Liebmann et al., 2004). In contrast, negative SIS phases are associated with an increased occurrence of intense and prolonged heat waves in the region (Cerne & Vera, 2011).

Between October and April, SIS activity on longer intraseasonal scales (SIS_LP) is significantly influenced by the MJO. Those long SIS events are often linked to intense MJO episodes with coherent eastward propagation (Alvarez et al., 2017). The positive phase of SIS_LP (i.e., wet and cold conditions in SESA, dry and warm conditions in the SACZ) typically occurs as the MJO convection moves along the equator from the eastern Indian Ocean to the western Pacific (MJO stages 3–6). The SIS_LP negative phase coincides with periods when the MJO convection is located between the eastern tropical Pacific and the western tropical Indian Ocean (MJO stages 7–2). The underlying physical mechanisms for this interaction involve the generation of Rossby waves from the equatorial western Pacific-eastern Indian Ocean region, which propagate poleward and induce regional circulation anomalies

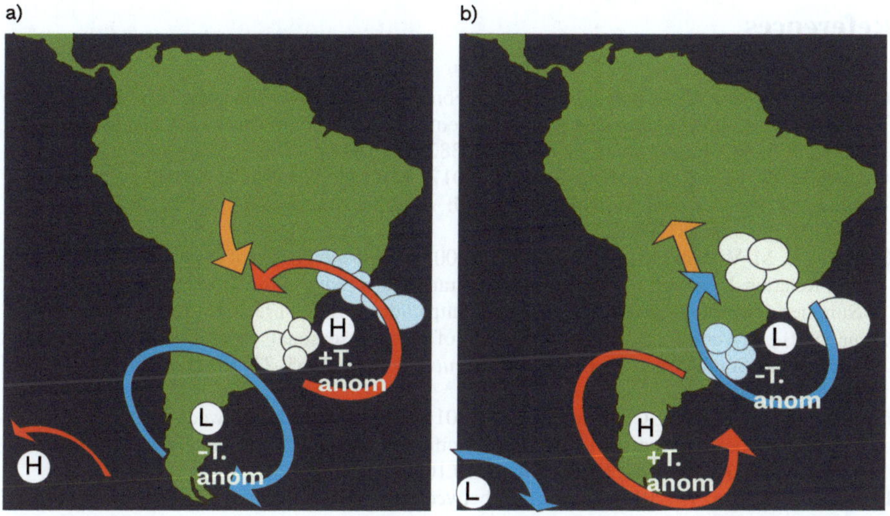

Fig. 3.4 Schematic of the regional anomalies associated with periods of (**a**) enhanced (white circles) and (**b**) inhibited convective cloudiness (light blue circles) over SESA during the warm season, linked to the positive and negative phases of the SIS pattern, respectively. The letters H and L, along with the corresponding circular arrows indicate anticyclonic and cyclonic circulation anomalies. +T and −T denote positive and negative temperature anomalies. The light orange arrow east of the central Andes indicates the direction of the low-level wind anomaly (Adapted from Vera et al. (2006))

over South America, combined with changes in the SACZ intensity driven directly by the MJO.

On shorter intraseasonal timescales, SIS activity (SIS_SP) during the same period (October to April) is also linked to wave trains crossing the South Pacific, although their origin is not yet fully understood. Potential sources include changes in convergence and divergence zones in the subpolar jet exit region south of Australia, interactions between convergence zones (such as the SACZ, the SPCZ, and the South Indian Convergence Zone - SICZ-, Vera et al., 2018), and nonlinear processes associated with equatorial waves (Vera et al., 2018; Raupp & Silva Dias, 2010).

During the cold season (May to September), when the South American Monsoon is inactive, the SIS spatial pattern is characterized by a monopole extending in a northwest-southeast direction over SESA on both short and long intraseasonal timescales, although the tilt is more pronounced for short timescales. Again, the pattern appears related to arched wave trains extending along the South Pacific (Vera et al., 2018). On longer time scales, the MJO is found to play a role in triggering the wave-trains that characterize the cold-season SIS_LP pattern, but the source of variability for the SIS_SP pattern remains uncertain. Interactions between the westerly flow and synoptic disturbances, as well as the influence of the SAM, may account for some of SIS_SP activity, though further research is needed to clarify these mechanisms.

References

Alvarez, M. S., Vera, C. S., Kiladis, G. N., & Liebmann, B. (2016). Influence of the Madden Julian Oscillation on precipitation and surface air temperature in South America. *Climate Dynamics, 46*(1), 245–262. https://doi.org/10.1007/s00382-015-2581-6

Alvarez, M. S., Vera, C. S., & Kiladis, G. N. (2017). MJO Modulating the Activity of the Leading Mode of Intraseasonal Variability in South America. *Atmosphere, 8*(12), 232. https://doi.org/10.3390/atmos8120232

Carvalho, L. M. V., Jones, C., & Ambrizzi, T. (2005). Opposite Phases of the Antarctic Oscillation and Relationships with Intraseasonal to Interannual Activity in the Tropics during the Austral Summer. *Journal of Climate, 18*, 702–718. https://doi.org/10.1175/JCLI-3284.1

Cerne, S. B., & Vera, C. S. (2011). Influence of the intraseasonal variability on heat waves in subtropical South America. *Climate Dynamics, 36*(11), 2265–2277. https://doi.org/10.1007/s00382-010-0812-4

Damião Mendes, M. C., & Cavalcanti, I. F. A. (2014). The relationship between the Antarctic oscillation and blocking events over the South Pacific and Atlantic Oceans. *International Journal of Climatology, 34*, 529–544. https://doi.org/10.1002/joc.3729

Flatau, M., & Kim, Y. (2013). Interaction between the MJO and polar circulations. *Journal of Climate, 26*, 3562–3574. https://doi.org/10.1175/JCLI-D-11-00508.1

Gonzalez, P. L. M., Vera, C. S., Liebmann, B., & Kiladis, G. (2008). Intraseasonal variability in subtropical South America as depicted by precipitation data. *Climate Dynamics, 30*(7), 727–744. https://doi.org/10.1007/s00382-007-0319-9

Gottschalck, J., Wheeler, M., Weickmann, K., et al. (2010). A Framework For Assessing Operational Madden–Julian Oscillation Forecasts. *Bulletin of the American Meteorological Society, 91*(9), 1247–1258. https://doi.org/10.1175/2010BAMS2816.1

Hirata, F. E., & Grimm, A. M. (2017). The role of synoptic and intraseasonal anomalies on the life cycle of rainfall extremes over South America: non-summer conditions. *Climate Dynamics, 49*, 313–326. https://doi.org/10.1007/s00382-016-3344-8

James, I. (1994) Introduction to circulating atmosphere. Cambridge University Press, United Kingdom

Kiladis, G. N., & Weickmann, K. M. (1992). Extratropical forcing of tropical Pacific convection during northern winter. *Monthly Weather Review, 120*(9), 1924–1938. https://doi.org/10.1175/1520-0493(1992)120<1924:EFOTPC>2.0.CO;2

Lau, W. K. M., & Waliser, D. E. (2011). *Intraseasonal variability of the atmosphere–Ocean–climate system*. Springer.

Liebmann, B., Kiladis, G. N., Marengo, J., et al. (1999). Submonthly convective variability over South America and the South Atlantic convergence zone. *Journal of Climate, 12*, 1877–1891. https://doi.org/10.1175/1520-0442(1999)012<1877:SCVOSA>2.0.CO;2

Liebmann, B., Kiladis, G. N., Vera, C. S., et al. (2004). Subseasonal variations of rainfall in South America in the vicinity of the low-level jet east of the Andes and comparison to those in the South Atlantic convergence zone. *Journal of Climate, 17*, 3829–3842. https://doi.org/10.1175/1520-0442(2004)017<3829:SVORIS>2.0.CO;2

Liu, F., Wang, B., Ouyang, Y., et al. (2022). Intraseasonal variability of global land monsoon precipitation and its recent trend. *npj Climate and Atmospheric Science, 5*, 30. https://doi.org/10.1038/s41612-022-00253-7

Lorenz, D. J., & Hartmann, D. L. (2001). Eddy-zonal flow feedback in the southern hemisphere. *Journal of the Atmospheric Sciences, 58*, 3312–3327. https://doi.org/10.1175/1520-0469(2001)058<3312:EZFFIT>2.0.CO;2

Madden, R. A., & Julian, P. R. (1972). Description of global-scale circulation cells in the tropics with a 40–50 day period. *Journal of the Atmospheric Sciences, 29*, 1109–1123. https://doi.org/10.1175/1520-0469(1972)029<1109:DOGSCC>2.0.CO;2

Mo, K. C., & Higgins, R. W. (1998). The Pacific-South American modes and tropical convection during the Southern Hemisphere Winter. *Monthly Weather Review, 126*, 1581–1596. https://doi.org/10.1175/1520-0493(1998)126<1581:TPSAMA>2.0.CO;2

Oliveira, F. N. M., Carvalho, L. M. V., & Ambrizzi, T. (2014). A new climatology for Southern Hemisphere blockings in the winter and the combined effect of ENSO and SAM phases. *International Journal of Climatology, 34*, 1676–1692. https://doi.org/10.1002/joc.3795

Raupp, C. F. M., & Silva Dias, P. L. (2010). Interaction of equatorial waves through resonance with the diurnal cycle of tropical heating. *Tellus A, 62*(5), 706–718. https://doi.org/10.1111/j.1600-0870.2010.00463.x

Renwick, J. A., & Revell, M. J. (1999). Blocking over the South Pacific and Rossby wave propagation. *Monthly Weather Review, 127*, 2233–2247. https://doi.org/10.1175/1520-0493(1999)127<2233:BOTSPA>2.0.CO;2

Thompson, D. W. J., & Wallace, J. M. (2000). Annular modes in the extratropical circulation. Part I: Month-to-month variability. *Journal of Climate, 13*, 1000–1016. https://doi.org/10.1175/1520-0442(2000)013<1000:AMITEC>2.0.CO;2

Trenberth, K. F., & Mo, K. C. (1985). Blocking in the Southern Hemisphere. *Monthly Weather Review, 113*, 3–21. https://doi.org/10.1175/1520-0493(1985)113<0003:BITSH>2.0.CO;2

Vera, C., Higgins, W., Amador, J., et al. (2006). Toward a unified view of the American monsoon systems. *Journal of Climate, 19*, 4977–5000. https://doi.org/10.1175/JCLI3896.1

Vera, C. S., Alvarez, M. S., Gonzalez, P. L. M., et al. (2018). Seasonal cycle of precipitation variability in South America on intraseasonal timescales. *Climate Dynamics, 51*, 1991–2001. https://doi.org/10.1007/s00382-017-3994-1

Wheeler, M. C., & Hendon, H. H. (2004). An all-season real-time multivariate MJO index: Development of an index for monitoring and prediction. *Monthly Weather Review, 132*, 1917–1932. https://doi.org/10.1175/1520-0493(2004)132<1917:AARMMI>2.0.CO;2

Zhang, C. (2005). Madden-Julian Oscillation. *Review of Geophysics, 43*, RG2003. https://doi.org/10.1029/2004RG000158

Chapter 4
Interannual Variability

4.1 Sources of Interannual Variability in South America

The climate in South America exhibits significant interannual variability across many of its regions (Vera & Silvestri, 2009). As discussed in Chap. 2, the standard deviation of seasonal means relative to the long-term or climatological average is a measure of the typical magnitude of interannual variability. Focusing on precipitation, during the summer (defined here as the January–February-March average), interannual variability is particularly large in the tropical regions of the continent, with maxima over the ITCZ and SACZ, and a secondary maximum over SESA (Fig. 4.1a). In autumn (defined as the April–May-June average), variability over SACZ nearly vanishes but remains very large in the ITCZ and notable over SESA (Fig. 4.1b). During winter (the July–August-September average), variability becomes negligible over most of the tropical continent, consistent with the dry season there, except the far north, but persists over SESA (Fig. 4.1c). In spring (the October–November-December average), variability reappears in the SACZ region, with the onset of the rainy season, and increases over the LPB (Fig. 4.1d). Thus, while interannual rainfall variability in the tropical LPB fluctuates seasonally in sync with the region's pronounced monsoon cycle, in the southern LPB this variability remains high throughout the year, often surpassing the amplitude of the seasonal cycle.

Previous studies have identified five factors that account for interannual variability in the tropical and subtropical regions of South America (Vera et al., 2006b):

1. **SST anomalies:** These can be local, such as those in the southwestern Atlantic, or remote, such as those in the tropical Pacific and Indian Oceans.
2. **Land surface condition anomalies:** Such as soil moisture or vegetation variations.

C. S. Vera et al., *Climate Variability and Change in Southern South America*, SpringerBriefs in Environmental Science,
https://doi.org/10.1007/978-3-032-07197-2_4

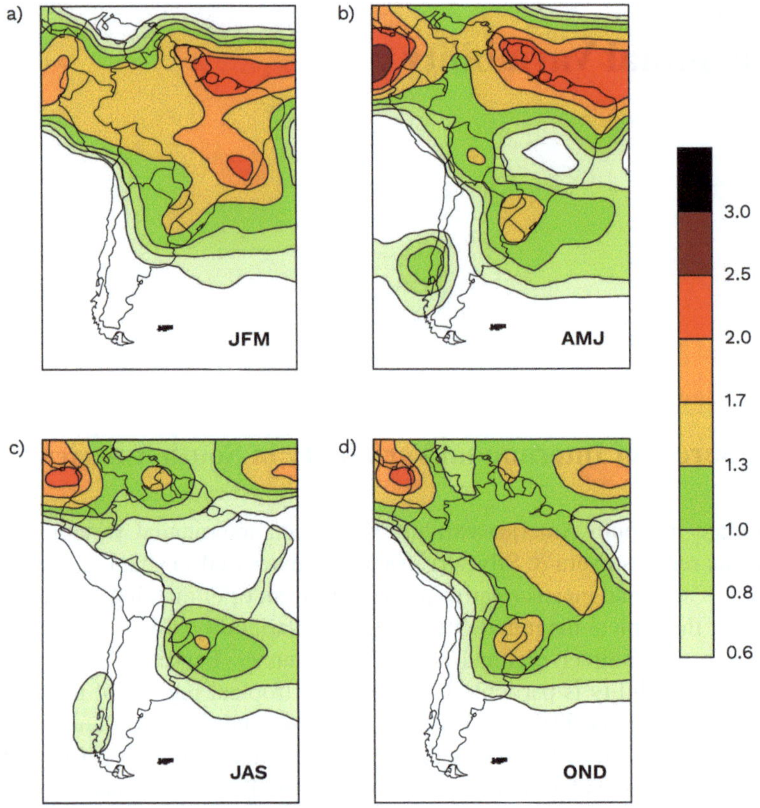

Fig. 4.1 Standard deviation of observed seasonal precipitation for (**a**) January–February-March, (**b**) April–May-June, (**c**) July–August-September, and d) October–November-December. Units: mm day^{-1} (Adapted from Vera and Silvestri (2009))

3. **Position and intensity of convergence zones:** Including the SACZ, the ITCZ, and even the remote SPCZ.
4. **Moisture transport anomalies:** Such as those associated with variations in the orientation, magnitude, and convergence of the SALLJ.
5. **Large-scale circulation anomalies:** Such as those associated with teleconnection patterns between the tropical Pacific and Indian Oceans and South America, including the PSA patterns, and the SAM.

The first three factors are strongly influenced by conditions external to the atmosphere and exhibit a degree of predictability. The last two factors instead are predominantly shaped by the internal variability of the atmosphere, making them less predictable. The following sections will discuss the most salient and relevant aspects of these factors and their influence on the interannual variability of southern South America and the LPB.

4.2 ENSO

ENSO is the most prominent manifestation of interannual variability in the coupled ocean-atmosphere system in the tropical Pacific Ocean. This basin is characterized by a strong longitudinal thermal gradient at the surface. While the monthly mean temperatures in the west, around Indonesia and north of Australia, can exceed 28 °C (the so-called "western Pacific warm pool"), the temperatures in the eastern Pacific and along the South American coast, can be lower than 20 °C. This pronounced east-west thermal contrast sustains a strong dynamic coupling between the ocean and the atmosphere (Fig. 4.2).

The differences in SST create a west-to-east pressure gradient that drives easterly surface winds along the equator. These winds converge in the west over the warm pool, causing vigorous upward air motion, condensation of water vapor, the formation of convective clouds, and precipitation. This process releases a large amount of latent heat in the mid-to-upper troposphere, providing energy that further accelerates the upward motion and the outflow at upper levels, from the warm region to the west (high pressure aloft) to the cold region to the east (low pressure aloft). The result is a zonal circulation cell along the equatorial Pacific, known as the Walker Cell, with a descending branch in the eastern part of the basin, off the South American coast, that reinforces the easterly surface winds. These interactions also

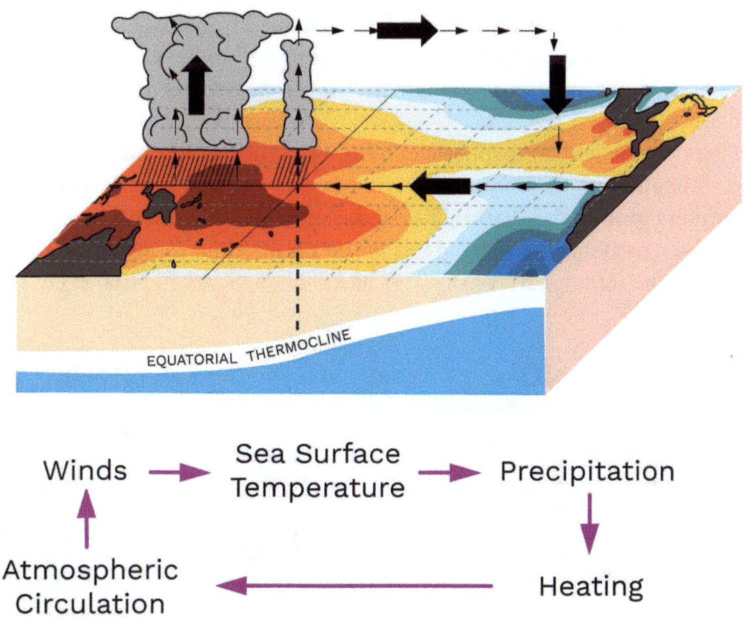

Fig. 4.2 Mean state of the ocean and atmosphere across the tropical Pacific during DJF including sea surface and subsurface temperatures (shaded), vertical and meridional circulation (arrows), and schematic convective clouds. Below, a diagram illustrates the links among the different variables involved in the air-ocean coupling (Adapted from NOAA/CPC ENSO site)

affect the oceanic surface boundary layer (via evaporation and upwelling of cold subsurface water in the east), making it deeper in the warm western end and shallower in the cooler eastern end. The coupling between the ocean and atmosphere is most intense during austral winter-spring and weaker during austral autumn.

ENSO, which is the leading pattern of variability in this system (e.g., Yang et al., 2018), arises from the same coupled ocean-atmosphere interactions. A full ENSO cycle comprises a "warm" phase, also known as "El Niño", and a "cold" phase, also referred to as "La Niña". During the El Niño phase, anomalously warm waters spread across the equatorial eastern Pacific, dampening the west-to-east pressure gradient and associated easterly winds and shifting the convection eastward from its climatological position (Fig. 4.3a). The weakened winds reinforce the warm SST anomalies through reduced upwelling and evaporation. In contrast, during the La Niña phase, the warm pool is confined to the far western Pacific, causing tropical convection to concentrate in that region, while anomalous subsidence and suppressed convection prevail over the central and eastern parts of the basin (Fig. 4.3b).

Both El Niño and La Niña events tend to develop between March and June, peak between December and April, and dissipate by May–July of the following year. El Niño episodes typically last between 9 and 12 months, and only occasionally longer, whereas La Niña events can persist more often for 1–3 years but are generally weaker in intensity.

ENSO influences the climate of many tropical regions primarily by altering the locations of upward or downward air motion. It also affects many subtropical, extratropical, and even polar regions in both hemispheres, by generating teleconnections via Rossby wave trains, which influence regional climate variability. Due to its extensive latitudinal range, South America is impacted by ENSO through both mechanisms.

Both the oceanic and atmospheric conditions in the tropics that can act as precursors and indicators of El Niño or La Niña events are monitored operationally by various centers around the world. One of the most historically used oceanic ENSO indices, known as Niño 3.4 (EN3.4), is based on the time series of SST anomalies averaged over the region 120°W - 170°W and 5°N - 5°S. An El Niño event is defined

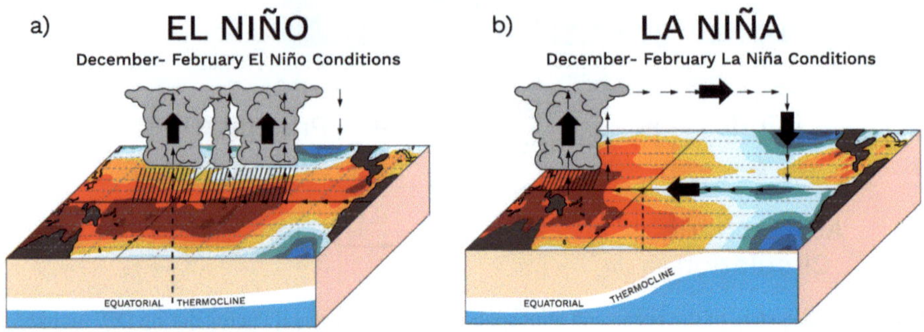

Fig. 4.3 Same as Fig. 4.2 but during (left) El Niño and (right) La Niña events (Adapted from NOAA/CPC ENSO site)

when the EN3.4 index persists above a certain threshold, while a sustained negative value indicates a La Niña event.

The impact of ENSO on a particular region can then be described by studying the statistical relationship between the variability of the EN3.4 index and that of the climatic anomalies in that region. Figure 4.4 shows "correlation maps", obtained by correlating the EN3.4 index with circulation anomalies at 500 hPa (the mid troposphere) and precipitation anomalies for each season (Vera & Silvestri, 2009). Circulation anomalies are represented by geopotential height anomalies, which are akin to pressure anomalies at upper levels (Sect. 2.1) and reflect the strength and pattern of the anomalous flow (positive values indicate anticyclonic or counterclockwise circulation in the Southern Hemisphere). The alternating anomalies of opposite sign along the South Pacific, South America, and as far as the southern Atlantic that occur simultaneously with ENSO events reflect teleconnections linking the tropical Indo-Pacific region with the high and mid-latitudes of the Southern Hemisphere (PSA-like patterns). The location, intensity, and sign of the anomalies in Fig. 4.4 then indicate the influence of an El Niño event on the climate of southern South America, with opposite effects during La Niña events. For example, the presence of a low-pressure cyclonic anomaly west of SESA and a high-pressure anticyclonic anomaly west of the Antarctic Peninsula in spring (OND) represents the teleconnection pattern most conducive to above-normal precipitation in SESA during El Niño events (Fig. 4.4, bottom panels).

Based solely on the EN3.4 index, the influence of El Niño on SESA, including the subtropical LPB, is strongest in the spring of the onset year and the autumn of the following year; in the interim seasons, the influence is negligible. However, it is now accepted that ENSO occurs in different "flavors", and thus the impacts of ENSO on South America climate may vary depending on the particular El Niño or La Niña type. For instance, in Eastern Pacific El Niño events SST anomalies in the equatorial Pacific peak near the South American coast, while in Central Pacific El Niño events anomalies maximize in the central part of the basin. These differences extend to the associated circulation and rainfall responses in South America (Cai et al., 2020). Eastern-Pacific El Niño events are associated with above-normal rainfall in SESA across all four seasons, with the strongest anomalies typically occurring from September to May. Central-Pacific El Niño events also lead to positive rainfall anomalies in SESA, particularly during spring (Cai et al., 2020). However, the SESA summertime rainfall anomalies linked to Eastern-Pacific El Niño events are approximately three times greater than those associated with Central-Pacific events. Conversely, while Eastern-Pacific La Niña events result in summer precipitation deficits in SESA, Central-Pacific La Niña events are linked to even more severe rainfall reductions but during the spring.

The influence of ENSO can also be detected in the variability of river discharges within the LPB. El Niño events lead to increased streamflow in major rivers of the basin, including the Paraná, Uruguay, Negro, and Paraguay, while La Niña events are associated with reduced flows (e.g., Robertson & Mechoso, 1998; Antico & Vuille, 2022). These reductions can result in extremely low water levels, as observed during the multi-year La Niña episodes between 2020 and 2023 (Rivera, 2024).

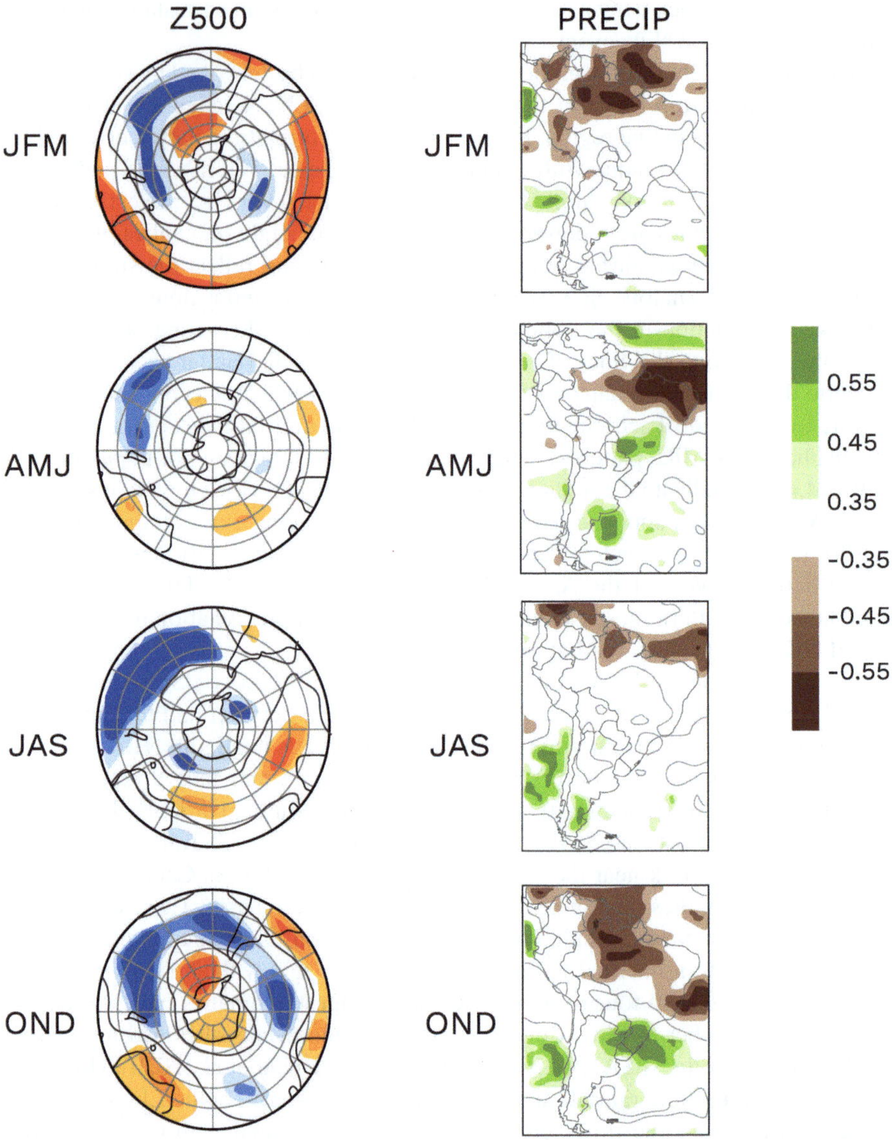

Fig. 4.4 Seasonal correlation between EN3.4 index and (left) circulation anomalies in the middle troposphere (represented by 500-hPa geopotential height anomalies), and (right) precipitation anomalies. The analysis was conducted for the period 1979–1999. Areas where positive (negative) correlations are statistically significant at 90, 95 and 99% based on a Student's T-test are shaded. In the panels on the left, values of 0.35, 0.45, and 0.55 are shown in yellow, orange, and red, while their negative counterparts are shown in light blue, blue, and dark blue. The zero contours are represented by a black line (Adapted from Vera and Silvestri (2009))

About two-thirds of the anomalously high discharges recorded in the Paraná River at Corrientes City in Argentina occurred during El Niño events, either in the spring of the onset year or in the autumn of the following year (Camilloni & Barros, 2003). In contrast, no unusually large discharges have been observed during La Niña events. A similar relationship exists between ENSO and extreme flow events in the Uruguay River, where 13 out of 18 flood events have occurred during El Niño and none during La Niña. However, for this river, there is no specific time of year when the ENSO influence is strongest. Furthermore, while ENSO-driven interannual flow variability dominates in the Uruguay and Negro rivers, long-term trends and decadal variability are more prominent in the Paraná and Paraguay rivers (Robertson & Mechoso, 1998).

4.3 SAM

The leading pattern of seasonal atmospheric circulation variability in the Southern Hemisphere is the SAM, introduced in Chap. 3. SAM exhibits considerable variability on interannual and longer time scales that affect the circulation in the Southern Hemisphere and southern South America. SAM structure, as described by seasonal anomalies (Fig. 4.5), is much more zonally symmetric during the austral summer, whereas in autumn and spring the Antarctic anomalies protrude into the midlatitude Pacific (Fogt & Marshall, 2020 and references therein).

Although the SAM primarily arises from the interaction between the mean westerly flow and extratropical transient disturbances, on seasonal and interannual timescales it can also be influenced by SST anomalies induced by ENSO and other anomalous tropical patterns. The relationship between SAM and ENSO is such that positive SAM phases occur more frequently during La Niña events, while negative SAM phases are more common during El Niño events. This relationship can strengthen or weaken on interdecadal timescales (e.g., Silvestri & Vera, 2009), as discussed later in Chap. 5. Additionally, from winter to spring, SAM activity can be influenced by stratospheric variability linked to the dynamics of the polar vortex, the low-pressure center over the South Pole that develops at altitudes above 15 km (e.g., Thompson et al., 2011).

The SAM activity strongly modulates the circulation over and around southern South America, inducing seasonal anomalies in various climate variables, such as rainfall (Fig. 4.5) and surface air temperature (e.g., Vera & Silvestri, 2009). Specifically, in the LPB region, the SAM influence on rainfall is significant in all seasons but summer. During autumn and spring, the relationship is negative, meaning that positive SAM phases are associated with below normal precipitation. In winter, the relationship reverses, with positive SAM phases linked to above-average rainfall (Fig. 4.5). Moreover, the combined influence of ENSO and SAM on precipitation in the LPB can be especially pronounced during the austral spring, often leading to unusually wetter conditions when El Niño events coincide with negative SAM phases (Vera & Silvestri, 2009). However, the SAM's influence on LPB

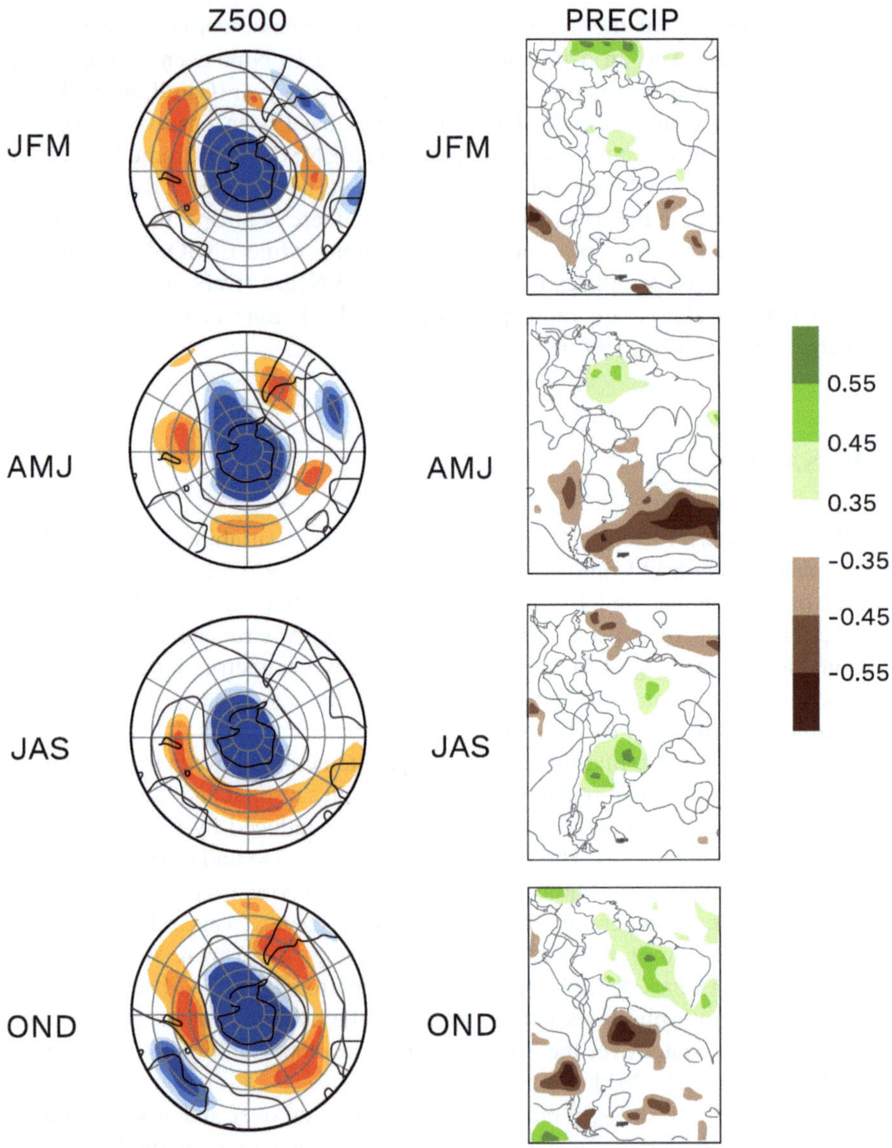

Fig. 4.5 As in Fig. 4.4, except fields are correlated simultaneously with a SAM index. (Adapted from Vera and Silvestri (2009))

rainfall anomalies is not particularly stable and can exhibit large variations from one decade to another (Silvestri & Vera, 2009), as discussed in Chap. 5.

4.4 Influence of the Indian and Atlantic Oceans

In addition to the tropical Pacific, changes in surface oceanic conditions in other tropical regions, such as the Indian and Atlantic oceans, can also give rise to climate variations in the LPB.

The leading mode of SST variability in the tropical Indian Ocean is known as the Indian Ocean Dipole (IOD). Its spatial structure consists of a west-east dipole of SST anomalies (Andrian et al., 2024 and references therein). In its positive phase, the IOD displays positive SST anomalies in the western tropical Indian Ocean and negative anomalies in the eastern part of the basin; the negative phase exhibits the opposite configuration. IOD events of significant magnitude typically develop around May or June, peak between August and October and dissipate rapidly by December, lasting between 2 and 7 months. The IOD exhibits a quasi-periodicity of 2–4 years, though its timing is highly irregular.

While the strongest IOD effects are observed in regions bordering the Indian Ocean, the IOD also triggers atmospheric circulation anomalies in the extratropics of the Southern Hemisphere, including southern South America, via teleconnections. During austral spring, the positive IOD phase is associated with significant above-normal precipitation in SESA, including the subtropical LPB, and below-normal precipitation in southeastern Brazil, where the tropical sector of the LPB lies (Sena & Magnusdottir, 2021). The teleconnection wavy pattern that causes this rainfall dipole is characterized by an anticyclonic anomaly south of South America and a subtropical cyclonic anomaly in the southeastern Pacific. The spring precipitation response to the negative phase of the IOD tends to be stronger in the subtropical LPB than in its tropical portion (Sena & Magnusdottir, 2021), although it is generally less consistent (Chan et al., 2008).

Although the IOD is an independent mode from ENSO, the two can overlap in time and, depending on their respective phases, reinforce or dampen the teleconnections they induce in the Southern Hemisphere. During austral spring, a positive IOD phase can amplify the circulation anomalies over the South Pacific triggered by an El Niño event, particularly in South America, resulting in stronger rainfall anomalies in SESA. However, this additive effect is not observed when La Niña events coincide with negative IOD phases (Andrian et al., 2024).

SST variability in the tropical Atlantic primarily influences northern and northeastern Brazil, through its interaction with the ITCZ (e.g., Mo & Berbery, 2011). Its effect on SESA is relatively weak compared to that of ENSO. However, like Indian SST anomalies, tropical Atlantic SST anomalies can either reinforce or counteract ENSO's impact on atmospheric circulation and rainfall over this part of South America (e.g., Martín-Gómez & Barreiro, 2016). Some studies suggest that the relative contribution of the tropical oceans (Pacific, Indian, and Atlantic) has changed over the past century. Using complex network analysis applied to observations, Martín-Gómez and Barreiro (2016) found that, around the 1930s, the combination of ENSO and tropical Atlantic variability was the dominant driver of precipitation variability in SESA. More recently, however, the joint influence of ENSO and the IOD has become more important (Martín-Gómez & Barreiro, 2016).

The variability of the adjacent South Atlantic does not seem to directly affect climate variability in SESA. However, numerical model experiments suggest that this oceanic region can modulate the influence of ENSO (Barreiro, 2010). During El Niño events, the anomalous atmospheric circulation tends to induce warming in the southwestern Atlantic during spring. This warming then modifies the regional circulation, ultimately impacting surface air temperature and precipitation in SESA (Barreiro, 2010).

4.5 Main Patterns of Regional Climate Variability

The leading patterns of variability in precipitation and other climate-related variables on seasonal and interannual scales can be identified using EOF analysis. According to Grimm (2011), the leading pattern of annual rainfall variability in South America exhibits a center of action over the subtropical LPB and another of opposite sign along the northern coast of the continent. This pattern is well correlated with SST anomalies in the Pacific and Indian Oceans that resemble those associated with ENSO and the IOD. That is, wetter conditions in the subtropical LPB described by this leading pattern are associated to positive SST anomalies in the central-eastern tropical Pacific and the western tropical Indian Ocean and to negative anomalies in the western tropical Pacific and the extratropical central Pacific (Grimm, 2011).

While the leading pattern of autumn rainfall variability resembles the annual one, the leading patterns in other seasons exhibit notable differences in their spatial distribution and associated SST influences. In winter, the dominant pattern is more localized, featuring a single stronger center of activity over the LPB; spring and summer display a different dipolar structure, with one center of action in the subtropical LPB and another of opposite sign in eastern and southeastern Brazil, where the SACZ develops. And while the SST anomalies associated with the leading spring pattern are strongly linked to both ENSO and IOD, in summer the ENSO influence is weaker.

The impact of the South American monsoon on rainfall variability in the LPB is also evident and significant throughout the year, except in the winter. During spring and summer, rainfall anomalies in the SACZ region tend to oscillate out of phase with those over the subtropical LPB. Negative rainfall anomalies in the LPB during these seasons are associated with local subsidence linked to increased convection in the SACZ region (e.g., Robertson & Mechoso, 2000).

The interannual and longer-term variability of river flows in the main sub-basins of the LPB is largely driven by precipitation variability (Gulizia & Camilloni, 2022).

In the tropical sub-basins (Paraná, Paraguay, and Pantanal), streamflow variability is strongly influenced by precipitation over the SACZ region. In contrast, the subtropical sub-basins of Uruguay and Iguazú rivers are more sensitive to year-to-year fluctuations in extratropical atmospheric circulation, which modulate the intensity of moisture transport into these regions.

Variability in land surface conditions, such as soil moisture and vegetation, can also affect atmospheric processes. This control, often referred to as "land-atmosphere coupling", is defined in literature based on the linear relationship between a slowly varying land variable, such as soil moisture, and a faster variable, such as precipitation (e.g., Spennemann & Saulo, 2015 and references therein). As one would expect, the influence of soil moisture on regional precipitation is more pronounced in summer. Additionally, land-atmosphere interactions can intensify and increase the persistence of extreme temperature events. While positive soil moisture–precipitation feedbacks (i.e., higher soil moisture leads to increased precipitation and vice-versa) have been identified in the SACZ region, negative feedbacks have been observed in the SESA region (Spennemann & Saulo, 2015).

References

Andrian, L. G., Osman, M., & Vera, C. S. (2024). The role of the Indian Ocean Dipole in modulating the austral spring ENSO teleconnection to the Southern Hemisphere. *Weather and Climate Dynamics, 5*, 1505–1522. https://doi.org/10.5194/wcd-5-1505-2024

Antico, A., & Vuille, M. (2022). ENSO and Paraná flow variability: Long-term changes in their connectivity. *International Journal of Climatology, 42*(14), 7269–7279. https://doi.org/10.1002/joc.7643

Barreiro, M. (2010). Influence of ENSO and the South Atlantic Ocean on climate predictability over southeastern South America. *Climate Dynamics, 35*, 1493–1508. https://doi.org/10.1007/s00382-009-0666-9

Cai, W., et al. (2020). Climate impacts of the El Niño–Southern Oscillation on South America. *Nature Reviews Earth & Environment, 1*, 215–231. https://doi.org/10.1038/s43017-020-0040-3

Camilloni, I. A., & Barros, V. R. (2003). Extreme discharge events in the Paraná River and their climate forcing. *Journal of Hydrology, 278*, 94–106. https://doi.org/10.1016/S0022-1694(03)00133-1

Chan, S. C., Behera, S. K., & Yamagata, T. (2008). Indian Ocean Dipole influence on South American rainfall. *Geophysical Research Letters, 35*(14), 10–14. https://doi.org/10.1029/2008GL034204

Fogt, R. L., & Marshall, G. J. (2020). The southern annular mode: Variability, trends, and climate impacts across the southern hemisphere. *WIREs Climate Change, 11*, e652. https://doi.org/10.1002/wcc.652

Grimm, A. M. (2011). Interannual climate variability in South America: Impacts on seasonal precipitation, extreme events, and possible effects of climate change. *Stochastic Environmental Research and Risk Assessment, 25*, 537–554. https://doi.org/10.1007/s00477-010-0420-1

Gulizia, C., & Camilloni, I. A. (2023). Relationship between rainfall and streamflow in the La Plata Basin: Annual cycles, interdecadal and multidecadal variability. *Atmosfera, 36*(2), 183–205. https://doi.org/10.20937/atm.53013

Martín-Gómez, V., & Barreiro, M. (2016). Analysis of oceans' influence on springtime rainfall variability over Southeastern South America during the 20th century. *International Journal of Climatology, 36*, 1344–1358. https://doi.org/10.1002/joc.4428

Mo, K. C., & Berbery, E. H. (2011). Drought and persistent wet spells over South America based on observations and the U.S. CLIVAR drought experiments. *Journal of Climate, 24*, 1801–1820. https://doi.org/10.1175/2010JCLI3874.1

Rivera, J. A. (2024). Characterization of the recent (2019–2022) La Plata Basin hydrological drought from a centennial-scale perspective. *HydroResearch, 7*, 140–153. https://doi.org/10.1016/j.hydres.2024.02.002

Robertson, A. W., & Mechoso, C. R. (1998). Interannual and decadal cycles in river flows of Southeastern South America. *Journal of Climate, 11*, 2570–2581. https://doi.org/10.1175/1520-0442(1998)011<2570:IADCIR>2.0.CO;2

Robertson, A. W., & Mechoso, C. R. (2000). Interannual and Interdecadal variability of the South Atlantic convergence zone. *Monthly Weather Review, 128*, 2947–2957. https://doi.org/10.1175/1520-0493(2000)128<2947:IAIVOT>2.0.CO;2

Ruscica, R. C., Sorensson, A. A., Diaz, L. B., et al. (2022). Evapotranspiration trends and variability in southeastern South America: The roles of land-cover change and precipitation variability. *International Journal of Climatology, 42*(4), 2019–2038. https://doi.org/10.1002/joc.7350

Sena, A. C. T., & Magnusdottir, G. (2021). Influence of the Indian Ocean dipole on the large-scale circulation in South America. *Journal of Climate, 34*, 6057–6068. https://doi.org/10.1175/JCLI-D-20-0669.1

Silvestri, G., & Vera, C. S. (2009). Nonstationary impacts of the southern annular mode on southern hemisphere climate. *Journal of Climate, 22*, 6142–6148. https://doi.org/10.1175/2009JCLI3036.1

Spennemann, P. C., & Saulo, A. C. (2015). An estimation of the land-atmosphere coupling strength in South America using the global land data assimilation system. *International Journal of Climatology, 35*, 4151–4166. https://doi.org/10.1002/joc.4274

Thompson, D. W. J., & Wallace, J. M. (2000). Annular modes in the extratropical circulation. Part I: Month-to-month variability. *Journal of Climate, 13*, 1000–1016. https://doi.org/10.1175/1520-0442(2000)013<1000:AMITEC>2.0.CO;2

Thompson, D. W. J., Solomon, S., Kushner, P., et al. (2011). Signatures of the Antarctic ozone hole in southern hemisphere surface climate change. *Nature Geoscience, 4*, 741–749. https://doi.org/10.1038/ngeo1296

Vera, C. S., & Silvestri, G. (2009). Precipitation interannual variability in South America from the WCRP-CMIP3 multi-model dataset. *Climate Dynamics, 32*, 1003–1014. https://doi.org/10.1007/s00382-009-0534-7

Vera, C., Higgins, W., Amador, J., et al. (2006). Toward a unified view of the American monsoon systems. *Journal of Climate, 19*, 4977–5000. https://doi.org/10.1175/JCLI3896.1

Yang, S., Li, Z., Yu, J.-Y., et al. (2018). El Niño–Southern Oscillation and its impact in the changing climate. *National Science Review, 5*(6), 840–857. https://doi.org/10.1093/nsr/nwy046

Chapter 5
Decadal Variability and Climate Change

5.1 Global and Regional Decadal Variability

There are two leading patterns of decadal variability driven by anomalous oceanic conditions that significantly influence the climate of South America: The Pacific Decadal Oscillation (PDO) and the Atlantic Multidecadal Oscillation (AMO) (Reboita et al., 2021 and references therein).

5.1.1 The Pacific Decadal Oscillation (PDO)

The PDO is the leading mode of decadal SST variability in the Pacific Ocean, primarily northward of 20°N (Mantua et al., 1997). Initially identified due to its effects on North American climate and its impact on marine and terrestrial ecosystems, subsequent research has revealed that the PDO exerts a broader influence on climate conditions well beyond this region.

Global climate centers typically describe the PDO as the leading mode of an EOF analysis applied to monthly SST anomalies in the North Pacific (north of 20°N), after removing both the annual cycle and the global-mean monthly anomaly. The resulting time series, known as the PDO index, is used to monitor the oscillation's behavior over time (Fig. 5.1c). Significant negative phases of the PDO occurred during 1890–1924, 1947–1976, and 2000–2010, while positive phases were observed in 1925–1946 and 1977–2000. More recently, the PDO shifted to predominantly positive values in the mid-2010s and returned to negative values after around 2020. Paleoclimatic reconstructions indicate that PDO variability extends at least as far back as 1600 (Mantua & Hare, 2002).

C. S. Vera et al., *Climate Variability and Change in Southern South America*, SpringerBriefs in Environmental Science,
https://doi.org/10.1007/978-3-032-07197-2_5

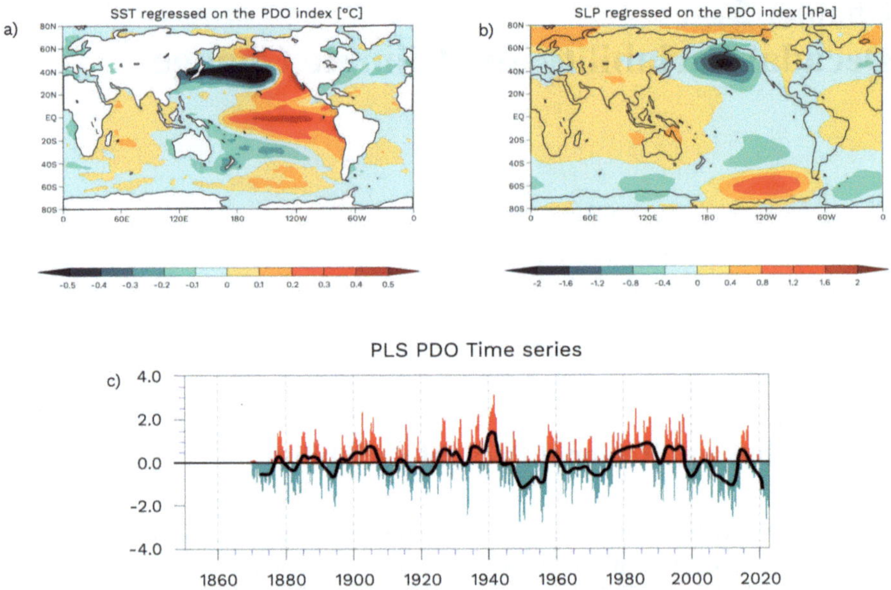

Fig. 5.1 (**a**) SST and (**b**) SLP anomaly patterns associated with the positive phase of the PDO, represented by the distribution of monthly anomalies linearly regressed onto the PDO index for one standard deviation of the index (adapted from the PDO site of Japan Meteorological Agency). (**c**) Time evolution of the normalized PDO index as described by monthly (shaded) and smoothed values (thick contour) (Adapted from the PDO site of NOAA/Physical Science Laboratory)

The positive phase of the PDO is characterized by strong negative SST anomalies in the central North Pacific, weaker anomalies in the South Pacific, and positive anomalies in the equatorial Pacific (Fig. 5.1a). In the Southern Hemisphere, the associated sea-level pressure pattern features weak negative anomalies in the central South Pacific and strong positive anomalies west of the Antarctic Peninsula (Fig. 5.1b), a configuration similar to the PSA pattern driven by El Niño (Fig. 4.4, Chap. 4). Conversely, the negative phase of the PDO exhibits negative SST anomalies in the equatorial Pacific and positive SST anomalies in the extratropical Pacific in both hemispheres. The anomalous SST patterns associated with the positive and negative phases of the PDO resemble those of El Niño and La Niña, respectively. However, while typical ENSO events last 6–18 months in the case of El Niño and 1–3 years for La Niña, PDO events typically persist for 20–30 years.

The climatic anomalies in SESA associated with the PDO are spatially similar to those linked to ENSO, but generally weaker in magnitude. The positive phase of the PDO is associated with warmer and wetter conditions in SESA, including the subtropical LPB, consistent with the low-pressure circulation induced over the subtropical South Pacific, which leads to an anomalous flux of tropical moist air from the northwest (Fig. 5.1b). Conversely, the negative phase is linked to drier and cooler conditions over the region (Reboita et al., 2021 and references therein).

Seasonally, under ENSO-neutral conditions, the PDO positive phase enhances precipitation over the LPB in summer (DJF) and, to a lesser extent, in spring (SON), while the negative phase is associated with drier conditions year-round especially in autumn (MAM). Importantly, the PDO can also modulate ENSO effects (Kayano & Andreoli, 2007). When El Niño coincides with the positive phase of the PDO, rainfall anomalies intensify across the entire LPB. However, when El Niño occurs during a negative PDO phase, positive precipitation anomalies concentrate in the subtropical LPB, while drier conditions prevail further north (Silva et al., 2011).

The streamflows of the Paraguay and Paraná rivers also exhibit pronounced variability on decadal timescales (Robertson & Mechoso, 1998; Maciel et al., 2013). The influence of the PDO has been particularly strong on the Paraná River's discharges, showing a significant positive correlation with the PDO index during summer (Perez et al., 2021).

5.1.2 Atlantic Multidecadal Oscillation (AMO)

The AMO is the leading pattern of decadal SST variability in the North Atlantic (Fig. 5.2). The positive (warm) phase is characterized by widespread positive SST anomalies across the North Atlantic (e.g., Schlesinger & Ramankutty, 1994; Alvarez-Garcia et al., 2008). The AMO index is typically derived from an EOF analysis of detrended SST anomalies in the North Atlantic (e.g., Alvarez-Garcia et al., 2008). Alternatively, it can be monitored through the average SST anomalies over that basin (Fig. 5.2, e.g., Schlesinger & Ramankutty, 1994). The AMO exhibits a periodicity of approximately 60–90 years, with observed warm phases during 1925–1960, and 1995–present, whereas cold phases occurred during 1905–1925, and 1970–1990 (e.g., Reboita et al., 2021 and references therein).

The AMO's influence is strongest in central-northern South America, particularly in spring and autumn, where its positive phase weakens moisture transport inland, leading to reduced precipitation (Reboita et al., 2021). The influence of the AMO is generally weaker on the LPB region. However, during negative AMO phases, SALLJ events are more pronounced along the eastern slopes of the Andes, enhancing moisture transport and precipitation over the LPB (Jones & Carvalho, 2018).

The AMO also modulates the precipitation anomalies induced by ENSO over South America. El Niño (La Niña) events that coincide with the cold (warm) phase of the AMO tend to produce larger and more coherent precipitation anomalies in SESA than when both modes are in phase (Kayano & Capistrano, 2014).

The AMO also influences the variability of the Paraná and Uruguay river discharges, exhibiting a negative correlation with summer discharge levels (Perez et al., 2021).

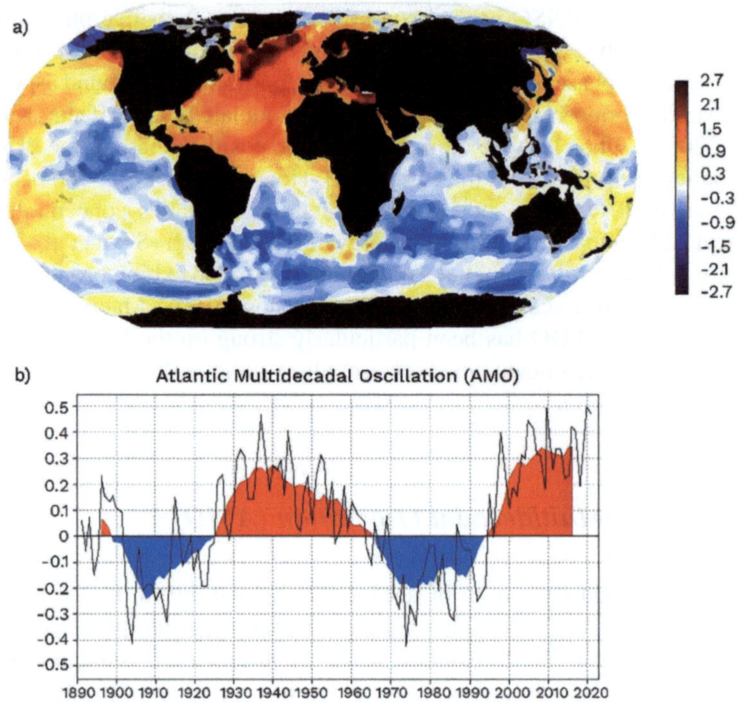

Fig. 5.2 (**a**) SST anomaly pattern associated with the positive phase of the AMO, represented by the distribution of monthly anomalies linearly regressed onto the AMO index for one standard deviation of the index (Adapted from Ruzmaikin, 2021). (**b**) Time evolution of the AMO index described by (black line) monthly and (colored) smoothed values. The AMO index is defined as the area-averaged SST over the North Atlantic, with the linear trend for the 1901–2000 removed (Adapted from the AMO site of Japan Meteorological Agency)

5.2 Climate Change Driven by Human Activities

5.2.1 Temperature

Since the beginning of the industrial era in the late nineteenth century, human activities have significantly altered the climate, mainly through increased emissions of greenhouse gases (GHGs), stratospheric ozone depletion, and changes in land use. While the first two processes cause alterations in the atmospheric chemical composition with global consequences, land-use changes primarily have localized and regional effects (Sect. 5.4).

In the mid-1970s, scientists detected severe depletion of stratospheric ozone due to the anthropogenic accumulation of halogen-containing gases (chlorine and bromine) in the atmosphere. A decade later, the "ozone hole" over Antarctica—where ozone depletion was the most extreme—was discovered. Since the adoption of the

1987 Montreal Protocol, efforts have been made to phase out the harmful ozone-destroying compounds. Evidence suggests that Antarctic stratospheric ozone likely reached its minimum levels around 2001 and has since slowly recovered, with projections indicating a return to normal levels between 2040 and 2060, depending on the region (e.g., Chipperfield & Bekki, 2024). The depletion and subsequent recovery of stratospheric ozone have had —and will continue to have— significant effects on climate conditions, particularly in the Southern Hemisphere's polar region, where these ozone-driven changes interact with those caused by rising GHG concentrations from human activities. While ozone depletion during the twentieth century led to cooling of the upper polar stratosphere and a strengthening of the westerly winds near 60°S, the post-2001 ozone recovery has already caused warming of the lower polar stratosphere and a weakening of the stratospheric polar vortex (e.g., Zambri et al., 2021).

Naturally present GHGs in the atmosphere—such as CO_2, water vapor, methane, and nitrous oxide–absorb part of the infrared radiation emitted by the Earth's surface, which is heated by incoming solar radiation, and then re-radiate infrared energy back toward the surface. In effect, the atmosphere provides the surface with an additional source of energy beyond the direct input from the Sun. This natural process, known as the greenhouse effect, results in surface air temperatures that are warm enough to support life on most of the planet. Human activities, however, have released large quantities of additional GHGs into the atmosphere, intensifying the greenhouse effect and causing an increase in global mean temperature (global warming), along with widespread changes in climate, including altered rainfall patterns, that vary by region and season. The main source of anthropogenic CO_2 emissions, the primary man-made GHG driving climate change, is the burning of fossil fuels (such as petroleum derivatives and coal). Other GHG gases, particularly methane and nitrous oxide, are primarily produced by agricultural and livestock activities.

According to the IPCC (2021) report, climate changes have been extensive and have occurred at an accelerated rate. By 2020, the global mean surface air temperature had risen by 1.1 °C relative to the 1850–1900 period (the beginning of the industrial era), reaching a level of warming unmatched over the past 2000 years. Superimposed on the long-term warming trend are decadal fluctuations associated with natural, large-scale patterns of multidecadal variability, such as the PDO and the AMO (Eyring et al., 2021).

Furthermore, recent decades have witnessed a global increase in the frequency and severity of heatwaves and heavy rainfall events, along with an increase in the occurrence of droughts and fire-prone weather conditions (warmer, drier, and windier) in many regions. These observed trends provide additional evidence of ongoing climate change.

Future climate simulations (termed "projections") presented in the IPCC report (2021) indicate that the climate changes that have already affected all regions of the planet in multiple ways will intensify as the mean global temperature continues to rise. Drawing from a wide range of studies, including meteorological records, paleoclimate reconstructions, satellite data and climate model simulations, IPCC (2021) has synthesized observed and projected changes in key *climate impact*

CID	FUTURE CHANGES	TREND / ATTRIBUTION
HEAT AND COLD		
Mean surfice temperature	∧ High confidence of increase	∧ Upward trend without attribution
Extreme heat	∧ High confidence of increase	∧ Upward trend with high confidence of attribution
Cold Spell	∨ High confidence of increase	∨ Downward trend with high confidence of attribution
Frost	∨ Medium confidence of increase	∨ Downward trend without attribution
WET AND DRY		
Mean precipitation	∧ High confidence of increase	∧ Upward trend without attribution
River Flood	∧ Medium confidence of increase	—
Heavy precipitation and pluvial flood	∧ Medium confidence of increase	∧ Upward trend without attribution

Fig. 5.3 Changes in the climate impact drivers synthetized by IPCC (2021) for SESA region. Future changes refer to a 20–30-year period centered around 2050 and/or consistent with 2 °C global warming, compared to a similar period within 1960–2014. Trend/attribution refers to the changes observed in the present climate. More information available at https://interactive-atlas.ipcc.ch/regional-synthesis/about/

drivers (CIDs) across many regions of the globe, including SESA (Fig. 5.3). In this region, mean surface air temperatures and extreme heatwaves have shown a clear increasing trend in recent decades, and there is high confidence that this warming will continue, as projected on average by climate models. Conversely, both cold waves and frost events have shown decreasing trends, with high and medium confidence, respectively, that they will continue to decline in the future.

5.2.2 *Precipitation*

Climate models are much less consistent when projecting changes in precipitation than in temperature, which stems from the fact that regional precipitation patterns are controlled not only by thermodynamic factors (e.g., temperature and moisture availability) but also by atmospheric circulation. In the SESA region, both annual and summer rainfall trends have been positive over recent decades, and it is projected that this trend will persist, based again on the ensemble of future climate simulations. The projections also indicate a continuation of the observed increase in extreme rainfall events and pluvial flooding. This is consistent with thermodynamic expectations based on the Clausius-Clapeyron equation, which states that the availability of moisture increases with increasing temperature (assuming constant relative humidity). Since the heaviest rainfall events occur when all the moisture in a volume of air is precipitated out, the intensity of extreme rainfall events should increase with temperature—provided that the atmospheric circulation does not change in ways that are very unfavorable to precipitation (i.e., a shift in the jet). When it comes to seasonal precipitation, however, uncertainties in circulation changes lead to uncertainties in both the sign and magnitude of precipitation changes. A recent study by Mindlin et al. (2023), highlights this uncertainty and complexity of regional precipitation responses, as different climate models produce plausible future scenarios for SESA featuring both increasing and decreasing precipitation trends.

Understanding these contrasting precipitation scenarios requires a closer examination of the large-scale atmospheric circulation patterns and remote forcings that influence the SESA climate. As discussed in previous chapters, the climate of SESA, which includes the subtropical LPB (Fig. 1.1), is influenced by large-scale phenomena of both tropical and polar origin. In the context of anthropogenic climate change, these phenomena are being—and will continue to be—altered by two major human-driven forcings: stratospheric ozone changes and rising GHG concentrations (Fig. 5.4, Mindlin et al., 2023). While ozone depletion and recovery mostly affect the activity of the stratospheric polar vortex, GHG increases modify not only the polar vortex but also tropical climate dynamics and the variability of large-scale patterns such as ENSO, the PDO, and the AMO. Changes in these patterns subsequently affect mid-latitude atmospheric circulation, including storm tracks, the Hadley circulation, and planetary wave activity, all of which can lead to regional circulation changes that either enhance or suppress precipitation in SESA. Moreover, the combined effects of ozone recovery and increasing GHG concentrations could drive a complex evolution of the climate in SESA over the twenty-first century, although the influence of ozone recovery appears to be secondary (Mindlin et al., 2021).

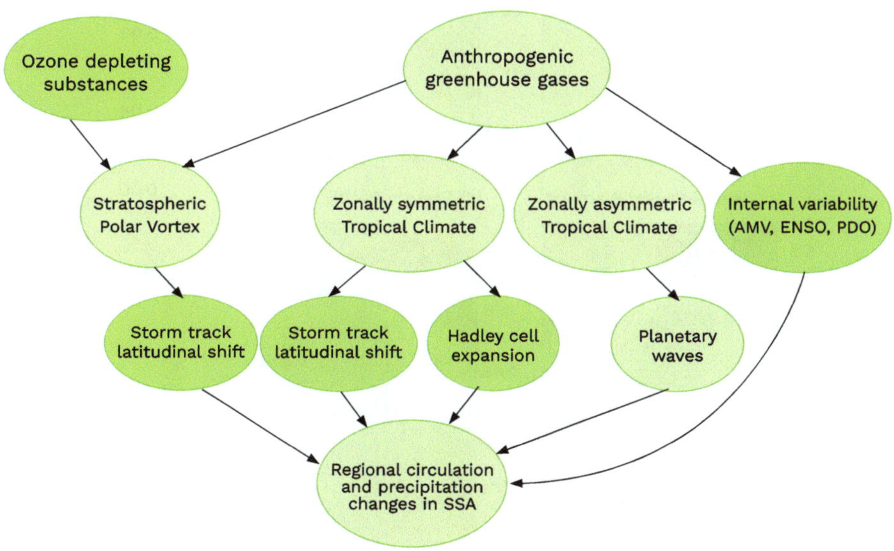

Fig. 5.4 Conceptual diagram of the remote drivers influencing circulation and precipitation changes over southern South America (adapted from Mindlin et al. (2023)). From top to bottom, the rows show (1) anthropogenic climate drivers, (2) large-scale phenomena, (3) elements of mid-latitude atmospheric circulation, and (4) regional changes. The darker shading indicates the elements explicitly considered in Mindlin et al. (2023)

By analyzing a large ensemble of climate projections from 28 different models, Mindlin et al. (2023) identified multiple possible future scenarios for summer rainfall changes in SESA due to rising GHGs. They found that four key remote drivers (RDs) will shape future climate conditions in the region: (1) the amplification of zonal mean tropical warming (or warming that is zonally uniform in the tropics), (2) the delayed breakup of the polar vortex (the persistence of strong stratospheric westerlies over Antarctica into late spring) and (3) two RDs related to positive SST changes in the central and eastern tropical Pacific (Fig. 5.4).

Each of these RDs has an effect on atmospheric circulation that can either enhance or suppress rainfall in SESA. For instance, zonal mean tropical warming promotes an expansion of the Hadley cell and a poleward shift of mid-latitude westerlies—trends already detected and expected to continue—resulting in subsidence conditions over SESA that suppress rainfall (Fig. 5.5a). Conversely, a delayed breakup of the polar vortex prolongs its influence on the troposphere, strengthening mid- and high-latitude westerly winds while enhancing easterly winds between

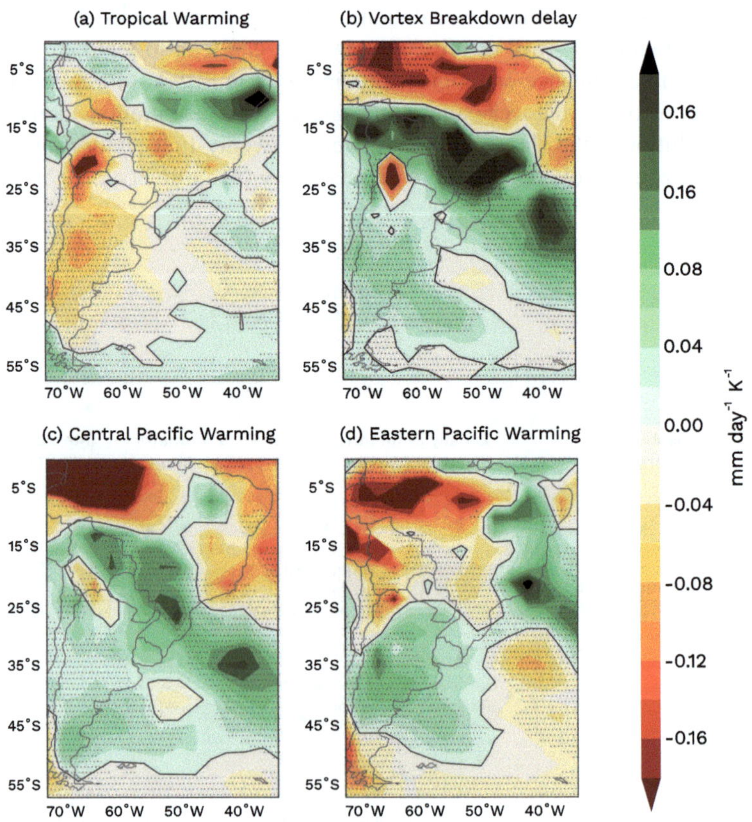

Fig. 5.5 Sensitivity of standardized summer precipitation to 4 remote drivers, described as the change associated with one standard deviation positive change in (**a**) tropical warming, (**b**) stratospheric polar vortex, (**c**) Central Pacific warming, and (**d**) Eastern Pacific warming. Stippling indicates areas with statistically significant values (Adapted from Mindlin et al. (2023))

30°S and 45°S over the eastern coast of South America, which in turn increases moisture transport into SESA (Fig. 5.5b). Finally, by decomposing projected changes in tropical Pacific SSTs into their zonal mean and asymmetric components, warming trends have been identified in both the central and eastern Pacific that resemble "El Niño" patterns and also modulate precipitation in SESA (Fig. 3 in Mindlin et al., 2023). These warming trends drive Southern Hemisphere circulation changes resembling PSA patterns that favor above-normal rainfall in SESA (Figs. 5.5 c and d). The key uncertainty, therefore, lies in how these four RDs will evolve and interact with each other in the future, and in the relative magnitude of both their changes and their effects, which will ultimately determine the sign and strength of precipitation changes in SESA.

Mindlin et al. (2023) analyzed these uncertainties using a novel "storylines" approach to describe a range of plausible future scenarios—providing insights that are often lost when focusing only on the ensemble-mean of climate projections. Most combinations of the four RDs identified for SESA suggest wetter conditions in the future. Specifically, both observed and projected rainfall increases in SESA are primarily associated with strong localized warming in the central and eastern Pacific, RDs that tend to dominate over zonal mean tropical warming and changes in the polar vortex. However, within these plausible future scenarios, it is also possible that a drier future could emerge if strong uniform tropical warming and early polar vortex breakup override the influence of central and eastern Pacific warming.

5.3 Combined Influence of Natural Climate Variability and Anthropogenic Trends in the LPB

As discussed previously, the combined influence of large-scale variability patterns (e.g., ENSO, SAM, IOD, PDO, AMO) has contributed to periods of both increased and decreased precipitation in SESA and fluctuations in river flows within the LPB. However, these interactions have exhibited multidecadal variations and, in some cases, have been further altered in recent decades by anthropogenic climate change.

The combined influence of ENSO and SAM in shaping climate anomalies over the LPB has varied considerably from decade to decade (Silvestri & Vera, 2009). While the relationship between ENSO and rainfall anomalies in SESA has been in general positive (i.e., El Niño events lead to increased rainfall, Fig. 4.4), the correlation between SAM and ENSO, as well as that between SAM and rainfall in SESA, has shifted over time (Silvestri & Vera, 2009). Before the 1980s, the correlation between SAM and rainfall in SESA during late spring was positive (with low pressures over Antarctica coinciding with above-average rainfall), but in more recent decades it has reversed to negative (Fig. 4.5). Additionally, while the most intense El Niño events recorded up to 2015 occurred in association with negative SAM events, the 2015–16 El Niño was accompanied by an exceptionally positive

SAM phase, leading to negligible rainfall anomalies in SESA during that summer (Vera & Osman, 2018). A similar situation occurred again in the summer of 2023–2024, when a strong El Niño event coincided with a persistently positive SAM phase, resulting in negative rainfall anomalies over SESA in January 2024, rather than an El Niño-induced increase. While the IPCC assessment (Eyring et al., 2021) does not provide conclusive evidence of anthropogenically-forced changes in the frequency or intensity of El Niño/La Niña events, it does report a significant trend toward positive SAM phases, driven by both rising GHG concentrations and stratospheric ozone depletion. Moreover, Mindlin et al. (2023) show that the interplay between human-induced changes in these two remote climate drivers (tropical warming associated with ENSO and polar forcing linked to SAM) could potentially lead to both wetter and drier than normal conditions in SESA in the present climate.

The interaction between the IOD and ENSO has also exhibited significant decadal variations, with a strong positive correlation between the 1970s and 1990s that has subsequently declined (Ham et al., 2016). Furthermore, the connection between negative IOD events and La Niña has weakened considerably since 1999 (Zu et al., 2024), and climate model simulations suggest that this weakening is likely to persist with higher levels of global warming (Zu et al., 2024). Conversely, it has been proposed that under global warming conditions, a new type of positive IOD event may emerge, potentially altering the relationship between positive IOD events and El Niño (e.g., Sun et al., 2022). As discussed in Chap. 4, precipitation in SESA is sensitive to both Indian and Pacific SST anomalies, particularly through their influence on the South American monsoon system and upper-level circulation. Accordingly, the interactions between ENSO and the IOD on decadal times scales are believed to be important for hydroclimate variability in the region. Quantifying and predicting the influence of these evolving ENSO-IOD interactions on rainfall variability in SESA remains an open question.

Similarly, the strength of ENSO's influence on SESA rainfall has varied on multidecadal timescales, in association with phase changes in the PDO and AMO. During the negative (cold) phase of the PDO that lasted approximately from 1941 to 1982, ENSO's teleconnection to SESA rainfall was weaker than between 1876–1940 and 1983–2016, and the correlation between ENSO and Paraná River flows also diminished (Antico & Vuille, 2022). In contrast, both precipitation in SESA and the Paraná River discharge exhibited their highest values between roughly 1970 and 2019 (e.g., Rivera, 2024), which has been attributed to the combined influence of a negative AMO phase, a positive PDO phase, and predominantly El Niño conditions (Perez et al., 2021 and references therein). Paleoclimate reconstructions based on sediment records from lakes and rivers within the LPB indicate that this wet period was the most intense in the past 230 years (Piovano et al., 2004).

Human influence, particularly rising GHG concentrations, has been linked to the observed increase in SESA rainfall in recent decades (e.g., Vera & Díaz, 2015). This influence has lead to a higher frequency of regional circulation anomalies conducive to above-normal summer precipitation (e.g., Díaz & Vera, 2017), consistent with the effect of planetary waves on regional precipitation described in the conceptual model displayed in Fig. 5.4.

Despite this positive rainfall trend since the mid-1950s, several severe droughts have occurred in the basin, with the most extreme and widespread event taking place between 2019 and 2023 (Rivera, 2024). The exceptionally low river levels observed during this period were unprecedented in the last 50 years of record and, in some areas of the southern basin, unmatched in the last 110 years. This extreme drought has been linked to two consecutive La Niña events during this period, combined with the anthropogenically-driven southward shift of the mid-latitude westerlies, which led to above-normal local temperatures and drier conditions in the LPB (Rivera, 2024 and references therein). This competing mechanism is consistent with the influence of storm-tracks shift and the Hadley cell on regional precipitation described in Fig. 5.4.

Using simulations with climate and hydrological models, Montroull et al. (2018) project that future warming will lead to increases in precipitation, evapotranspiration, and river discharge in the LPB's major rivers. However, due to the opposing effects on different components of the regional hydrological cycle and their associated mechanisms (such as those described in Fig. 5.4), considerable uncertainty remains regarding how these changes will interact in the future and their ultimate impact on river flows.

5.4 Impact of Land Use Change on Regional Climate

The hydrological behavior of the LPB is shaped not only by natural climate variability and anthropogenic climate change driven by rising GHG emissions, but also by the wide range of human activities carried out within the region (e.g., Perez et al., 2021). As noted in the introduction, the basin is densely populated, with major urban centers and varied economic sectors. More than half of South America's Gross National Product is generated within this basin. Large areas, particularly in southern Brazil, Uruguay, eastern Argentina, and Paraguay, are used for agriculture, affecting carbon sequestration, soil conservation, and water quality (Perez et al., 2021 and references therein).

The substantial land-use changes that have taken place across the basin, particularly since the second half of the twentieth century—and even earlier in some regions of Argentina and Brazil—have influenced not only river flows but also other key components of the regional hydrological cycle. Ruscica et al. (2022) examined changes in evapotranspiration across the basin and their relationship to both climate variability and land use changes. Evapotranspiration, which refers to the loss of water from moist soils and plant surfaces to the atmosphere, primarily depends on precipitation and solar radiation, while also being conditioned by surface properties such as soil moisture and vegetation cover. Land-use practices influence the energy and water balance by modulating evapotranspiration through biophysical (e.g., surface reflectivity or albedo) and biogeochemical processes (e.g., changes in CO_2

emissions from agricultural activities). Additionally, soil erosion caused by human activities can impact evapotranspiration. Ruscica et al. (2022) evaluated the combined effects of local land-use changes and precipitation variability and trends on regional evapotranspiration. Their analysis used simulations from global dynamic vegetation models run under varying atmospheric conditions (described by three different atmospheric gridded datasets) and considering both natural vegetation and observed land-use changes across southern South America from 1950 to 2010. Their findings revealed a positive trend in evapotranspiration in the southern portion of the basin, driven by precipitation variability remotely influenced by tropical Pacific forcing such as ENSO. In contrast, in the northern portion of the basin, significant land use changes induced a negative trend in evapotranspiration.

To assess the impact of surface process changes on the atmosphere, models capable of simulating land-atmosphere coupling are required. In this context, the term "coupling" refers to the control that a surface variable, such as soil moisture, can exert on an atmospheric variable (e.g., precipitation), considered independently from the feedback of the atmospheric variable on soil moisture (e.g., Ruscica et al., 2015). Regions where land-atmosphere coupling is particularly strong are known as "hot spots." The SESA region and eastern Brazil have been widely recognized as such hot spots. The influence of surface conditions in these regions extends beyond local effects, impacting atmospheric variables such as pressure, temperature and moisture in surrounding areas (Giles et al., 2022). For example, negative soil moisture anomalies in eastern Brazil have been shown to enhance moisture transport from the tropics toward that region, at the expense of SESA (Giles et al., 2022 and references therein).

References

Alvarez-Garcia, F., Latif, M., & Biastoch, A. (2008). On multidecadal and quasi-decadal North Atlantic variability. *Journal of Climate, 21*, 3433–3452. https://doi.org/10.1175/2007JCLI1800.1

Antico, A., & Vuille, M. (2022). ENSO and Paraná flow variability: Long-term changes in their connectivity. *International Journal of Climatology, 42*(14), 7269–7279. https://doi.org/10.1002/joc.7643

Chipperfield, M. P., & Bekki, S. (2024). Opinion: Stratospheric ozone—Depletion, recovery and new challenges. *Atmospheric Chemistry and Physics, 24*, 2783–2802. https://doi.org/10.5194/acp-24-2783-2024

Díaz, L. B., & Vera, C. S. (2017). Austral summer precipitation interannual variability and trends over Southeastern South America in CMIP5 models. *International Journal of Climatology, 37*, 681–695. https://doi.org/10.1002/joc.5031

Eyring, V., Gillett, N. P., Achuta Rao, K. M., et al. (2021). Human influence on the climate system. In V. Masson-Delmotte, P. Zhai, A. Pirani, et al. (Eds.), *Climate change 2021: The physical science basis. Contribution of working group I to the sixth assessment report of the intergovernmental panel on climate change* (pp. 423–552). Cambridge University Press. https://doi.org/10.1017/9781009157896.005

Giles, J. A., Menéndez, C. G., & Ruscica, R. C. (2022). Nonlocal impacts of soil moisture variability in South America: Linking two land–atmosphere coupling hot spots. *Journal of Climate, 36*, 227–242. https://doi.org/10.1175/JCLI-D-21-0510.1

Ham, Y.-G., Choi, J.-Y., & Kug, J.-S. (2016). The weakening of the ENSO–Indian Ocean Dipole (IOD) coupling strength in recent decades. *Climate Dynamics, 49*(1–2), 249–261. https://doi.org/10.1007/s00382-016-3339-5

IPCC. (2021). Summary for policymakers. In V. Masson-Delmotte, P. Zhai, A. Pirani, et al. (Eds.), *Climate change 2021: The physical science basis. Contribution of working group I to the sixth assessment report of the intergovernmental panel on climate change*. Cambridge University Press. https://doi.org/10.1017/9781009157896

Jones, C., & Carvalho, L. M. V. (2018). The influence of the Atlantic multidecadal oscillation on the eastern Andes low-level jet and precipitation in South America. *npj Climate and Atmospheric Science, 1*, 40. https://doi.org/10.1038/s41612-018-0050-8

Kayano, M. T., & Andreoli, R. V. (2007). Relations of South American summer rainfall interannual variations with the Pacific Decadal Oscillation. *International Journal of Climatology, 27*, 531–540. https://doi.org/10.1002/joc.1417

Kayano, M. T., & Capistrano, V. B. (2014). How the Atlantic multidecadal oscillation (AMO) modifies the ENSO influence on the South American rainfall. *International Journal of Climatology, 34*, 162–178. https://doi.org/10.1002/joc.3674

Maciel, F., Díaz, A., & Terra, R. (2013). Multi-annual variability of streamflow in La Plata Basin. Part I: Observations and links to global climate. *International Journal of River Basin Management, 11*(4), 345–360. https://doi.org/10.1080/15715124.2013.847843

Mantua, N. J., & Hare, S. R. (2002). The Pacific decadal oscillation. *Journal of Oceanography, 58*, 35–44. https://doi.org/10.1023/A:1015820616384

Mantua, N. J., Hare, S. R., Zhang, Y., et al. (1997). A Pacific interdecadal climate oscillation with impacts on salmon production. *Bulletin of the American Meteorological Society, 78*, 1069–1080. https://doi.org/10.1175/1520-0477(1997)078<1069:APICOW>2.0.CO;2

Mindlin, J., Shepherd, T. G., Vera, C. S., & Osman, M. (2021). Combined effects of global warming and ozone depletion/recovery on southern hemisphere atmospheric circulation and regional precipitation. *Geophysical Research Letters, 48*(12). https://doi.org/10.1029/2021GL092568

Mindlin, J., Vera, C. S., Shepherd, T. G., & Osman, M. (2023). Plausible drying and wetting scenarios for summer in southeastern South America. *Journal of Climate, 36*, 7973–7991. https://doi.org/10.1175/JCLI-D-23-0134.1

Montroull, N. B., Saurral, R. I., & Camilloni, I. A. (2018). Hydrological impacts in La Plata basin under 1.5, 2 and 3 °C global warming above the pre-industrial level. *International Journal of Climatology, 38*, 3355–3368. https://doi.org/10.1002/joc.5505

Perez, L., Barreiro, M., Etchevers, I., et al. (2021). Centennial hydroclimatic and anthropogenic processes of South East South America modulate interannual and decadal river discharge. *Science of the Total Environment, 781*, 146733. https://doi.org/10.1016/j.scitotenv.2021.146733

Piovano, E. L., Ariztegui, D., Bernasconi, S. M., & Mckenzie, J. A. (2004). Stable isotopic record of hydrological changes in subtropical laguna Mar Chiquita (Argentina) over the last 230 years. *Holocene, 14*(4), 525–535. https://doi.org/10.1191/0959683604hl729rp

Reboita, M. S., Ambrizzi, T., Crespo, N., et al. (2021). Impacts of teleconnection patterns on South America climate. *Annals of the New York Academy of Sciences, 1504*, 116–153. https://doi.org/10.1111/nyas.14592

Robertson, A. W., Mechoso, C. R. (1998). Interannual and Decadal Cycles in River Flows of Southeastern South America. *J Climate 11*:2570–2581. https://doi.org/10.1175/1520-0442(1998)0112570:IADCIR2.0.CO;2.

Rivera, J. A. (2024). Characterization of the recent (2019–2022) La Plata Basin hydrological drought from a centennial-scale perspective. *HydroResearch, 7*, 140–153. https://doi.org/10.1016/j.hydres.2024.02.002

Ruscica, R. C., Sorensson, A. A., & Menéndez, C. G. (2015). Pathways between soil moisture and precipitation in southeastern South America. *Atmospheric Science Letters, 16*, 267–272. https://doi.org/10.1002/asl2.552

Ruscica, R. C., Sorensson, A. A., Diaz, L.B., et al (2022) Evapotranspiration trends and variability in southeastern South America: The roles of land-cover change and precipitation variability. *International Journal of Climatology. 42*(4):2019–2038. https://doi.org/10.1002/joc.7350

Ruzmaikin, A. (2021). Climate patterns: Origin and forcing. *American Journal of Climate Change, 10*, 204–236. https://doi.org/10.4236/ajcc.2021.102010

Schlesinger, M. E., & Ramankutty, N. (1994). An oscillation in the global climate system of period 65–70 years. *Nature, 367*, 723–726. https://doi.org/10.1038/367723a0

Silva, G. A. M., Drumond, A., & Ambrizzi, T. (2011). The impact of El Niño on South American summer climate during different phases of the Pacific Decadal Oscillation. *Theoretical and Applied Climatology, 106*, 307–319.

Silvestri, G., & Vera, C. S. (2009). Nonstationary impacts of the southern annular mode on southern hemisphere climate. *Journal of Climate, 22*, 6142–6148. https://doi.org/10.1175/2009 JCLI3036.1

Sun, S., Fang, Y., Zu, Y., et al. (2022). Increased occurrences of early Indian Ocean Dipole under global warming. *Science Advances, 8*, 1–9. https://doi.org/10.1126/sciadv.add6025

Vera, C. S., & Díaz, L. (2015). Anthropogenic influence on summer precipitation trends over South America in CMIP5 models. *International Journal of Climatology, 35*, 3172–3177. https://doi.org/10.1002/joc.4153

Vera, C. S., & Osman, M. (2018). Activity of the southern annular mode during 2015–2016 El Niño event and its impact on southern hemisphere climate anomalies. *International Journal of Climatology, 38*, e1288–e1295. https://doi.org/10.1002/joc.5419

Zambri, B., Solomon, S., Thompson, D. W. J., et al. (2021). Emergence of Southern Hemisphere stratospheric circulation changes in response to ozone recovery. *Nature Geoscience, 14*, 638–644. https://doi.org/10.1038/s41561-021-00803-3

Zu, Y., Sun, S., Fang, Y., et al. (2024). Weakening of La Niña impact on negative Indian Ocean Dipole under global warming. *Geophysical Research Letters, 51*, e2024GL110039. https://doi.org/10.1029/2024GL110039

Chapter 6
Regional Climate Services

6.1 General Concepts of Climate Services

Climate information that reflects current conditions and anticipates their evolution over weeks, seasons, or even years is essential to reduce the impact of adverse weather and climate conditions, while also identifying new opportunities and benefits for society. This information is crucial for climate-sensitive socioeconomic sectors to make informed decisions and effectively manage their activities. In 2009, the WMO convened the Third World Climate Conference, bringing together national meteorological and hydrological services (NMHS) from around the world, as well as representatives from the private sector and civil society, to jointly discuss how to develop, implement, communicate, and utilize climate services (WMO, 2011). Climate services are defined as the provision of climate information to support decision-making, which requires effective engagement between users and providers, the integration of useful and credible scientific information and knowledge, and the establishment of efficient mechanisms to ensure access to information tailored to users' needs (Hewitt et al., 2012). The challenge set forth by the WMO has driven improvements and expansion of climate observation systems, development of new monitoring and prediction tools, and identification of innovative approaches to communication and engagement with key stakeholders in climate-sensitive sectors.

The implementation plan of the Global Framework for Climate Services (GFCS), developed by the WMO (e.g., Hewitt et al., 2012), outlines four fundamental components for the successful deployment of climate services: (1) user-provider engagement (including a user-provider interface platform); (2) climate services information systems; (3) observation and monitoring tools; and (4) research, modeling, and prediction, with capacity-building efforts integrated into all four components. The Interface Platform organizes the information and facilitates interaction among decision-makers, researchers, and information providers. To achieve this, it relies on a climate services information system that operationally collects, stores, and

C. S. Vera et al., *Climate Variability and Change in Southern South America*, SpringerBriefs in Environmental Science,
https://doi.org/10.1007/978-3-032-07197-2_6

processes past and future climate data, generating products that support decision-making in climate-sensitive socioeconomic sectors. The Observations and Monitoring component, specifically, gathers, manages, and disseminates climate observations and related products, providing insights into both current conditions and recent climate evolution. The Research, Modeling, and Prediction component strengthens the system by developing products that provide future climate outlooks and enhance understanding of climate variability and change. The GFCS prioritizes the development and provision of services in five key areas that offer opportunities to improve human security and well-being and are emphasized in national adaptation plans (WMO, 2011). These priority areas are agriculture and food security, disaster risk reduction, health, water, and energy.

The GFCS has been implemented across different regions worldwide, yielding diverse approaches and outcomes (e.g., Vaughan et al., 2017). The initiative quickly gained traction in southern South America. The new tasks, however, introduced significant challenges for institutions such as the meteorological and hydrological services of the region, which often face constraints in terms of human and financial resources. These limitations hinder their ability to strengthen and expand regional monitoring and prediction strategies and to translate data into valuable climate information for decision-makers. Despite these challenges, Southern South America has made notable progress in implementing climate services, primarily through international and regional projects based on collaborative networks. These networks include NMHS, academic and research institutions, sectoral organizations, local private sector actors, other government institutions, and civil society. After more than a decade of effort, the region has accumulated successful experiences and valuable lessons. Vera (2024) reflects on this process by examining regional and national case studies aimed at delivering climate services to key socioeconomic sectors. This study provides the foundation for the discussion presented in the following sections.

6.2 Regional Climate Monitoring

The development and provision of climate services at regional and local scales require an information system that integrates multiple spatial and temporal scales. The Climate Services Information System, as defined by the WMO within the Global Framework for Climate Services (GFCS), operates at three levels: national, regional, and global (Fig. 6.1; e.g., Hewitt et al., 2021). Regarding the data necessary for climate monitoring, the system assumes that continuous observations, data rescue, and data management, including quality control, are conducted at the national level. At the regional level, these observations are integrated, while at the global level, additional data sources are incorporated, such as those derived from satellite data and reanalysis datasets.

The implementation of a climate services information system highlighted the critical need, particularly in southern South America, to integrate meteorological observations from different countries in order to develop regional climate

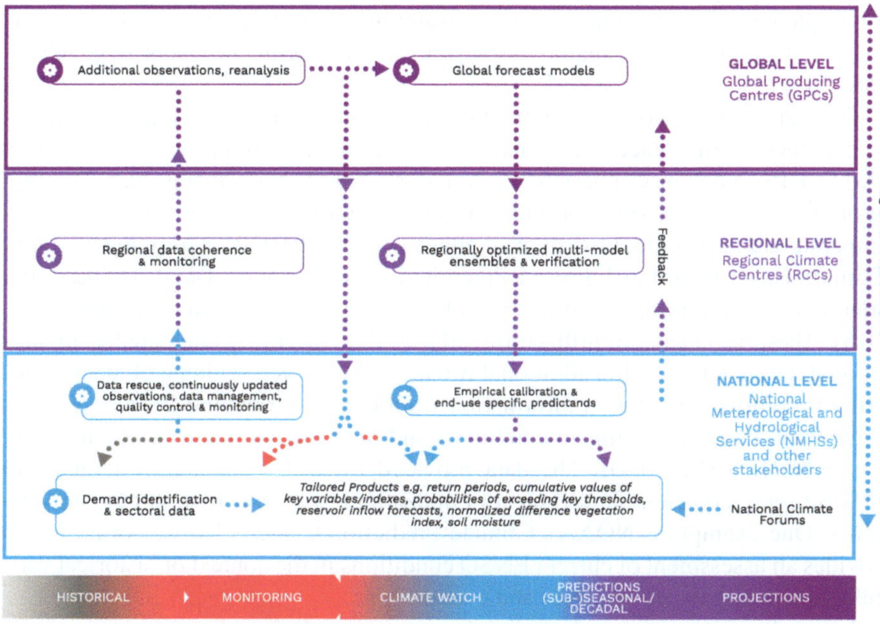

Fig. 6.1 Schematic of a regional approach to implement the Climate Services Information System (Adapted from Hewitt et al. (2021))

monitoring products. This necessity was identified by the GFCS (WMO, 2011), which also acknowledged the challenges that developing and underdeveloped countries face in implementing climate services. As a result, the WMO (2011) strongly recommended the establishment of regional climate centers to support the activities of national climate services. In response, the Regional Climate Center for Southern South America (CRC-SAS) was created as a virtual entity comprising the NMHS of Argentina, Brazil, Bolivia, Chile, Paraguay, and Uruguay, with the objective of strengthening and expanding climate service capabilities in the region.

Between 2012 and 2018, the CRC-SAS engaged with users, researchers, and representatives of sectoral institutions as part of an international project funded by the Inter-American Institute for Global Change Research (IAI) (Carabajal & Hidalgo, 2022). One of its first major achievements was the creation of the Regional Daily Database (RDDB), which has been operational since its inception. The CRC-SAS system routinely collects, conducts quality control on, and integrates daily meteorological observations from the six participating countries. The geographic coverage includes conventional and automatic weather stations across the continental region south of 10°S. The database currently includes daily maximum and minimum temperature records, as well as accumulated precipitation data provided by the six countries, with the potential for additional variables to be incorporated in the future (CRC-SAS, 2023). Despite the value of this database, the sparse spatial distribution of surface stations across the region has required the integration of complementary data sources, such as satellite-derived products and reanalysis datasets, to enhance the spatial resolution of monitoring products (e.g., CRC-SAS, 2024).

The characteristics of climate monitoring activities vary depending on their target audience. For instance, the primary users of climate monitoring products developed by CRC-SAS for southern South America are the NMHS themselves. Through its online portal, CRC-SAS provides a range of monitoring products focused on temperature and precipitation, derived either from the RDDB or from the CHIRPS (Climate Hazards Group InfraRed Precipitation with Station Data) database; the latter merges satellite-based estimates with surface station observations (Funk et al., 2015). These products are utilized directly by the NMHS or integrated with subnational datasets to develop additional products. Moreover, regional climate monitoring is complemented with information from Global Producing Centers (GPCs), which track large-scale climate conditions and their leading patterns of variability (e.g., ENSO, SAM, MJO). The data provided by GPCs are mainly targeted to NMHS users and scientific-technical organizations with expertise in climate variability. One example is NOAA's Climate Prediction Center's ENSO portal, which provides an assessment of current ENSO conditions in the context of historical variability, including expert analyses and evaluations.

A complementary approach to climate monitoring, designed to expand the accessibility and application of climate information in decision-making, focuses on specific climate hazards. These systems, which the WMO (2022) defines as "Climate Watch Systems," provide both real-time monitoring and predictive insights on timescales ranging from days to weeks and even seasons, specifically addressing climate conditions with potentially negative impacts. This information, which can be supported by data provided by a Regional Climate Center, is used by NMHS to issue warning messages. Climate Watch Systems focus on extreme meteorological and climate events such as heatwaves, cold waves, extreme precipitation, and droughts.

An example of such an initiative in southern South America is the Drought Information System for Southern South America (SISSA, by its Spanish acronym), which was launched in 2019 by CRC-SAS (Vera, 2024 and references therein). SISSA aims to reduce the social, economic, and environmental impacts of droughts in southern South America by generating and providing drought-related information and implementing actions to strengthen institutional capacities, planning, emergency preparedness, and governance in drought risk management. Since its inception, SISSA has engaged extensively with government agencies, academic institutions, the private sector, and civil society to co-develop and disseminate tailored monitoring products designed for diverse applications. Figure 6.2 illustrates an example of a SISSA monitoring product, which integrates data from four different meteorological drought indices derived from the RDDB. These indices are presented through text and graphics to depict the severity and evolution of drought events. Furthermore, SISSA, recognizing the detrimental effects not only of rainfall deficits (meteorological drought) but also of soil and vegetation moisture shortages (agricultural drought) and low river flows (hydrological drought), has developed

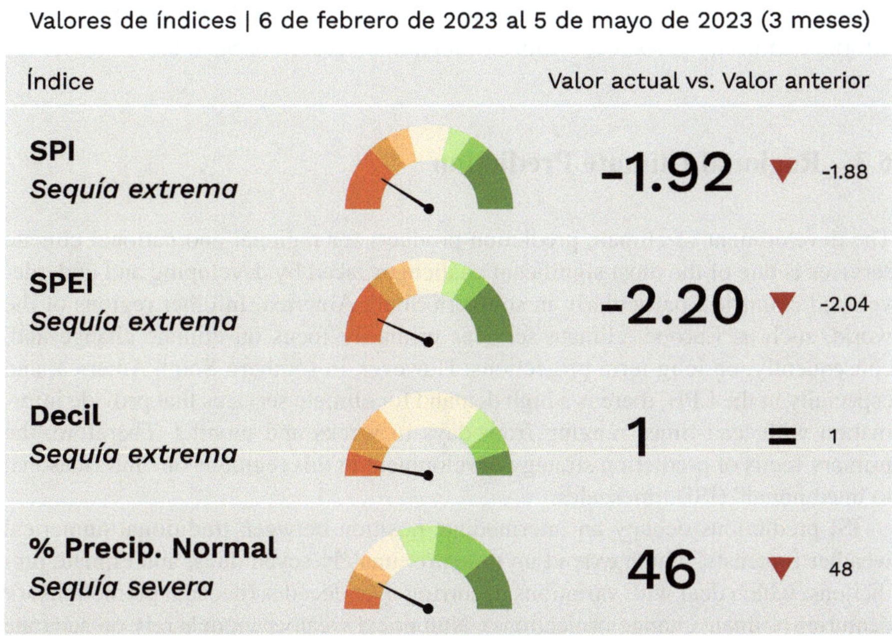

Fig. 6.2 Example of drought index products developed by SISSA (in Spanish) for Rocha, Uruguay, for a time scale of 6 months. The indexes considered are *SPI* (Standard Precipitation Index), *SPEI* (Standardized Precipitation-Evapotranspiration Index), *Decil* (Deciles of precipitation), and *% Precip. Normal* (Percentage of Normal Precipitation). The first column indicates the current category of conditions: dry, normal, or wet for each index. The second column indicates in graphic form the current value of each index (represented by the "needle" in the graph). The different categories of conditions (moderate to extreme or normal) are represented with different colors. The third column compares the current value (indicated by a large number on the left of the column) with the immediately preceding value (indicated by the smaller number to the right of the column) (Adapted from Vera (2024))

additional monitoring products that track soil conditions, vegetation status, and river levels at critical navigation points. These products have emerged from the intense collaboration the initiative has fostered with the agricultural, hydroenergy, and river navigation sectors.

At the national level, countries across the LPB have implemented various strategies to address climate hazards such as droughts. In Argentina, a national drought committee compiles and analyzes monthly information, not only on meteorological and climatic conditions but also on vulnerability and impacts—particularly in the agricultural sector—to assess and issue drought alerts for affected areas. Similarly, Chile and Brazil have established national platforms that continuously monitor drought severity based on various indicators and evaluate both short- and long-term impacts, serving as decision-support tools for different user groups. Additionally, several countries in the region are developing monitoring systems or products for other climate hazards, including heatwaves, floods, and wildfires, at different stages of implementation. Climate monitoring can also be tailored to meet the specific information needs of socioeconomic sectors, as discussed in Sect. 6.5.

6.3 Regional Climate Prediction

The development of climate prediction products for regional and national climate services is one of the most significant challenges faced by developing and underdeveloped countries, particularly in southern South America. In other regions of the world, such as Europe, climate services primarily focus on climate change and, consequently, on long-term projections. However, in southern South America, and especially in the LPB, there is a high demand for climate services that provide information with lead times ranging from days to weeks and months. Therefore, the primary focus of prediction strategy development in this region is on "intraseasonal to interannual" (ISI) timescales.

ISI predictions occupy an intermediate position between traditional numerical weather forecasts, which extend up to approximately seven days, and climate predictions, which deal with variations occurring over decades (decadal predictions) or centuries (climate change projections). Numerical weather models rely on accurate knowledge of initial atmospheric conditions to skillfully predict variations in atmospheric variables of interest for effective decision-making. In contrast, numerical climate models used for decadal prediction require only information on the current state of the slowly evolving components of the climate system, such as the oceans, cryosphere, and land surface. Even longer-term climate projections require only estimates of future GHG concentrations and other anthropogenic forcings. Bridging these two approaches, climate prediction on intermediate ISI timescales requires both initial atmospheric conditions and surface (land/ocean/ice) conditions (e.g., NRC, 2010).

ISI prediction skill depends both on the model's ability to represent climate-related mechanisms and on the intrinsic levels of climate predictability. Numerical climate models still present limitations in capturing certain physical processes—such as cloud-related processes, ocean-atmosphere interactions, and land-atmosphere coupling—and are constrained by the spatial resolution at which they operate, which affects forecast quality. However, even with "perfect" climate modeling systems, ISI predictions will never be flawless, as much of climate variability is inherently chaotic. Nonetheless, even if the predictable component is small it still provides valuable information. A key strategy to quantify and reduce the epistemic/structural uncertainty and the aleatoric uncertainty is the use of prediction ensembles using different models, combined with output from statistical methods (e.g., NRC, 2010).

The levels of climate predictability depend on the presence of predictability sources. In the case of the climate of southern South America, variability in the tropical oceans—primarily through ENSO and IOD activity—as well as variability in the stratospheric polar vortex and large-scale climate patterns such as the MJO, provide sources of predictability that make ISI predictions viable. The development of regional climate prediction tools requires an understanding of predictability levels and the quality of climate variability predictions. Potential predictability refers to the portion of total variability that can be predicted (the signal), with the remainder being the unpredictable component (the noise); it is a number between 0 and 1, often estimated through so-called "perfect-model experiments", in which the model's own simulated output is treated as the "truth" to be predicted. Potential predictability is always greater than actual prediction skill, which is mathematically assessed by correlating predictions with observations and can therefore be negative (but always smaller than 1).

The study of climate predictability over South America began in the early 2000s, although the region received less attention than others worldwide. Early assessments, based solely on global atmospheric models, showed moderate levels of predictability in SESA, but were limited in their ability to evaluate the role of air–ocean coupling—such as that involved in ENSO—in shaping predictability (Osman & Vera, 2017, and references therein). Osman and Vera (2017) assessed regional predictability on seasonal timescales using a set of retrospective forecasts based on past data to test model performance. Their analysis used nine different global atmosphere–ocean coupled models from the Climate Historical Forecast Project (CHFP); all were initialized one month before the target season. Their results indicated that, while both predictability and forecast skill for precipitation and surface air temperature were higher in tropical regions than in the extratropical areas of South America, the SESA region exhibited modest but statistically significant values. Furthermore, predictability for precipitation in SESA improved during ENSO years. For instance, the potential predictability of DJF precipitation in the LPB averaged between 0.5 and 0.6 but increased to between 0.6 and 0.7 in ENSO years. Although these levels are modest compared to those in tropical areas, they allow for high-value predictions. Similarly, de Andrade et al. (2019) assessed the quality of subseasonal rainfall predictions in SESA using eleven climate models from the Subseasonal to Seasonal Project (S2S). All models demonstrated good predictive skill for week 1, with performance declining in subsequent weeks. They also found that ENSO and the MJO provide potential sources of predictability for the region, enabling reasonable skill up to week 2.

Encouraged by these findings on seasonal predictability, efforts to develop regional climate prediction tools for SESA began during the 2000s and 2010s. Several meteorological services in the region utilized the Climate Prediction Tool (CPT), developed by the International Research Institute for Climate and Society (IRI; Mason, 2011), to produce the first prediction products. These tools were primarily based on the statistical relationship between SST variability in tropical regions and rainfall and surface air temperature anomalies in SESA, integrating observational data with outputs from global models available at the GPCs (e.g., Herrera, 2022; Campos et al., 2025). Since then, the "Centro de Previsão de Tempo e Estudos Climáticos" (CPTEC), in Brazil, has even developed and operates its own global climate model for both climate predictions and projections (Escada et al.,

2021 and references therein) — the only institution in South America currently doing so.

On the other hand, the availability of multi-model ensembles of global predictions, such as those from the North American Multi-Model Ensemble (NMME), has allowed the development of probabilistic seasonal forecasts for rainfall and temperature in the region (e.g., Coelho et al., 2006). These ensemble forecasts, generated for upcoming seasons, are further calibrated using local observational data to improve their reliability by reducing errors or biases relative to observed climate variability (Osman et al., 2021). By estimating the likelihood of specific climate conditions, probabilistic forecasts provide valuable guidance for decision-making and risk management. In particular, the prediction product developed by Osman et al. (2021) is operational at the CRC-SAS, delivering probabilistic forecasts for specific anomalous thresholds of key climate variables. These thresholds can also be customized based on user requirements. For example, the SISSA system generates probabilistic predictions of extreme dry or wet conditions, as well as extremely warm or cold conditions, to support impact assessments for phenomena such as droughts or floods.

Additionally, recent advances in understanding climate variability and predictability on subseasonal timescales have facilitated the development of subseasonal prediction systems, which are now available at certain GPCs globally and at CPTEC regionally. Using the Brazilian Global Atmospheric Model (BAM v1.2), Coelho et al. (2021) have implemented a subseasonal prediction capability at CPTEC that forecasts anomalies in a set of atmospheric variables for weeks 1–4, subseasonal prediction products tailored to specific user needs have begun to emerge, such as the BERMEJO product (Diaz et al., 2024). This tool, co-designed with agricultural stakeholders in northern Argentina, uses subseasonal forecasts from a global model to estimate the probability of different accumulated rainfall thresholds occurring in the second forecast week. The product has been evaluated through both traditional forecast verification methods and usability assessments by local stakeholders, demonstrating high potential (Diaz et al., 2024).

Other emerging prediction strategies in the region are "impact-based forecasts" (IBFs). These integrate meteorological and climate forecasts with data on vulnerability and exposure to specific risks affecting people and their livelihoods (Vera, 2024 and references therein). Creating IBFs requires new inter-institutional collaborations to collect and incorporate relevant data. The National Early Warning and Monitoring Center for Natural Disasters of Brazil (CEMADEN) has made significant progress in IBF development to support risk management for floods, flash floods, and mass movements (Moraes, 2023). The system combines probabilistic flood occurrence forecasts with impact assessments (derived from vulnerability and risk characterizations), translating them into actionable risk levels. Initial evaluations in Brazil indicate successful outcomes, particularly when alert levels are high.

6.4 Approaches to Developing and Disseminating Climate Information

One of the greatest challenges in providing climate services in a region like SESA is achieving meaningful engagement with relevant stakeholders who will use scientific knowledge, information, and products for decision-making (Vera, 2024). This need has driven the expansion and strengthening of interdisciplinary research that bridges climate science with fields such as agriculture, hydrology, health, and others, including the social sciences and humanities. Moreover, the traditional linear transfer model, in which knowledge flows directly from the academic-scientific-technical sector to decision-makers and users, has proven insufficient in ensuring that climate information is effectively used or valued by its recipients. Lemos et al. (2012) identified three key factors that create barriers to the usability of climate information: fit, interplay, and interaction. Fit includes issues like low spatial or temporal resolution and poor timing. Interplay refers to the user's limited technical capacity, lack of resources, or institutional constraints. Interaction involves weak communication, lack of trust, and limited collaboration between information producers and users. As an alternative, various conceptual models of engagement have been proposed in literature, some of which have been successfully applied in southern South America including the LPB, yielding promising results, as described below.

The initial implementation of the CRC-SAS was carried out within the framework of a joint proposal developed by climate information providers and users from regional countries (Carabajal & Hidalgo, 2022). This initiative created opportunities for institutional representatives and decision-makers to actively participate in defining the objectives, goals, and actions needed to establish effective climate services for the region. These partnerships strengthened collaboration among operational, governmental, scientific, and user communities, fostering new avenues for joint efforts. It is worth mentioning that a working structure on climate issues with such close involvement among agencies from different countries as the CRC-SAS did not previously exist in the region.

Hernandez et al. (2022) proposed a methodological framework based on knowledge co-production to develop climate services. This approach encourages interaction among diverse stakeholders—such as researchers, technical experts, local government officials, and actors from the agriculture and disaster risk management sectors, enabling them to integrate their knowledge and jointly generate new insights to inform the development of climate services. The authors then applied this approach to develop climate services tailored for family farming in northeastern Argentina by implementing "co-production cycles." In these cycles, stakeholders collaboratively identify climate-related challenges and propose potential solutions in the form of climate services. Research and development efforts then lead to a prototype service, which is tested and evaluated by users. Based on user feedback, the service either undergoes additional refinement cycles or proceeds to full implementation. A key example of this approach is the BERMEJO subseasonal prediction tool, mentioned in the previous section, along with the co-production of a mobile application that facilitates access to both this specific product and other co-produced monitoring tools.

As part of SISSA, demonstration projects have been implemented to co-develop climate services specifically designed to address challenges in the river navigation, hydroenergy, and agricultural sectors of the region (Vera et al., 2024). In the case of river navigation, stakeholders from the shipping and port sectors identified the need for a tool to predict water levels in the Paraná and Paraguay rivers at critical points for navigation. In response, researchers and technical experts created a forecasting tool based on a hydraulic model, calibrated with local observations and validated by users. This tool is openly and freely accessible through the SISSA portal.

6.5 Climate Information Tailored to Socioeconomic Sectors

As climate services in southern South America have continued to develop, the demand for more specific weather and climate knowledge tailored to the requirements of different socioeconomic sectors has increased. This has led not only to diversification in stakeholder engagement strategies for generating sector-specific climate services but also to the challenge of integrating climate information with non-climate data relevant to each sector. Vera (2024) reflects on various sectoral experiences across South America, highlighting that the most significant advancements in the region have occurred in climate services designed for the agriculture, water, and energy sectors, which we briefly discuss below.

Climate services tailored to the agricultural sector have been developed for various situations across the region through Decision Support Systems (DSS), which utilize observed and forecasted climate information to inform crop and farm models. These systems typically involve multiple institutions related to climate and agriculture and, in many cases, have been created through participatory or co-production processes. For example, in Uruguay, SIMAGRI ("Simulador de Agricultura") is a DSS co-designed to translate climate information into agricultural and economic terms, supporting decision-making for three key crops: corn, soybeans, and wheat (Han et al., 2019). Agricultural management decisions that benefit from such a system include selecting the type and variety of crops to plant, agricultural water management, planning for weed or disease control, fertilizer application, and yield insurance. This system integrates observed and forecasted climate data with information on soils and plant varieties, allowing users to run agricultural models to compare different management options under specific climate conditions, aiding decision-making both before and during the crop cycle. SIMAGRI is part of Uruguay's National Agricultural Information System, which provides support for decision-making in the sector. Loboguerrero Rodriguez et al. (2017) illustrate how this platform proved crucial in 2015, when the Uruguayan government declared an agricultural emergency in certain regions and implemented special credit lines to assist producers affected by a severe drought. Despite its demonstrated usefulness, one of the major challenges for DSS in LPB countries is their long-term continuity, as they are often developed within limited project lifespans, and regional, national, or local institutions may lack the capacity to sustain their ongoing operations.

The water and energy sectors have also been targeted in the LPB for the development of sector-specific climate services. Hydrological prediction systems have been increasingly developed across different time scales, ranging from daily to subseasonal and seasonal forecasts. This has introduced the challenge of not only assessing current hydrological conditions but also integrating climate predictions with hydrological models. In the LPB, there is a long-standing practice of using purely statistical approaches for streamflow forecasts of its major rivers. More recently, deterministic hydrological-hydrodynamic models have been employed, driven by climate outlooks derived either from historical observations projected into the future or from climate model outputs. For example, Petry et al. (2023) used a seasonal hydrological prediction system (SEAS5-SSF) to develop discharge forecasts of up to seven months for major South American rivers. This system employs the MGB-SA ("Modelo de Grandes Bacias-South America") hydrological-hydrodynamic model, which is initiated either by calibrated seasonal predictions from the ECMWF's fifth generation seasonal forecast system or by a set of future precipitation scenarios inferred from historical observations. Their findings indicate that hydrological predictions based on seasonal climate forecasts perform best for a one-month lead time for several rivers, including the Paraná, while those derived from precipitation-only scenarios are most reliable for lead times beyond two months.

Hydrological predictions designed to support the operation of hydroelectric plants have also received considerable attention in LPB countries. For instance, Kolling Neto et al. (2023) utilized the MGB-SA model, driven by medium-range outlooks from ECMWF, to produce weekly streamflow predictions for 147 hydroelectric plants in Brazil. Specifically, they assessed the impact of correcting climate forecasts with local observations and found that this adjustment significantly improved the quality of the predictions. Another noteworthy example comes from Uruguay, where several climate services have been developed to assist in decision-making related to the operation of the National Interconnected System, which manages the dispatch of electrical energy across the country (Terra & Baethgen, 2019). Specifically, the University of Uruguay and the Electricity Market Administration co-produced an optimization algorithm for the Electric Power System Simulator (SimSEE), designed to minimize direct costs and expected future expenses. Their findings indicate that incorporating ENSO predictions significantly reduced uncertainties and enhanced the planning of the electrical network, ultimately lowering its total operating costs (Maciel et al., 2015). More recently, the inclusion of 15-day hydrological forecasts from the Global Ensemble Forecast System model from NOAA in SimSEE has also demonstrated the potential to further improve its predictive accuracy (De Vera et al., 2022).

6.6 Future Perspectives

The development of climate services in the LPB, led by public institutions in collaboration with key sectoral stakeholders and users, has made significant progress over the past two decades. The success of regional collaboration efforts between public institutions, the private sector, national and regional organizations, and stakeholders is evident, enabling the implementation of initiatives such as the Regional Climate Center for Southern South America (CRC-SAS) and the Drought Information System for Southern South America (SISSA). These efforts support the work of national and subnational organizations such as NHMSs and sectoral institutions, providing a variety of tools and monitoring and prediction products for climate services. Additionally, climate services tailored to the specific needs of domains such as agriculture and water management (including hydro-energy) are expanding with promising results, for example, the SIMAGRI initiative of Uruguay (Han et al., 2019).

However, as described by Vera (2024), obstacles and challenges remain in the region, limiting the further expansion and strengthening of regional climate services. The region still faces significant gaps in its in-situ observation networks for atmospheric and surface conditions (including soil moisture and water resources), which cannot be fully compensated for by satellite data. Furthermore, the region lacks sufficient data on social, economic, and political conditions, among others, which are essential for comprehensively assessing the risks associated with climate hazards and mitigating their impacts.

To be sure, institutions in the region have advanced in the development and operational provision of climate monitoring and prediction products, particularly at seasonal timescales, with promising prospects for weekly to subseasonal forecasts. Yet, difficulties persist in accurately simulating key sources of climate variability in the LPB, such as land-atmosphere coupling processes (Ruscica et al., 2022). Moreover, probabilistic seasonal predictions, such as those developed by the CRC-SAS, would be especially valuable for countries in the region that currently lack the capacity to produce them independently. Overall, further progress is needed to generate climate predictions that are both locally relevant and actionable for decision-making in climate-sensitive sectors. Experiences in the agricultural and hydrological domains indicate clear opportunities for advancements in this regard. There is also a growing need to develop more IBFs. The experience with the IBF system for floods in Brazil demonstrates that improving the estimation of vulnerability among exposed populations is crucial for advancing these capabilities.

Several experiences in the region, such as SISSA, confirm that participatory and co-produced climate services have been both effective and promising. However, these initiatives have typically been limited in scope and duration, addressing specific issues within time-bound projects. Currently, most NMHS struggle, or are even unable, to sustain these efforts once project funding ends. There is a pressing need to replicate and scale up successful experiences while strengthening local and regional institutions to ensure the long-term sustainability of climate services.

Furthermore, fostering further development of climate services by private companies would be both desirable and necessary, as it would complement and diversify the services currently provided by the public sector.

References

Campos, D. A., Cabello, F. I., & Muñoz, A. G. (2025). On the NextGen-Chile forecast system: A calibrated multi-model ensemble approach for seasonal precipitation forecasts. *International Journal of Climatology, 45*, e8747. https://doi.org/10.1002/joc.8747

Carabajal, M. I., & Hidalgo, C. (2022). Making sense of climate science: From climate knowledge to decision-making. In P. Sillitoe (Ed.), *The Anthropocene of weather and climate: Ethnographic contributions to the climate change debate*. Berghahn Books. https://doi.org/10.2307/j.ctv29sfz69.12

Coelho, C. A. S., Stephenson, D. B., Balmaseda, M., et al. (2006). Toward an integrated seasonal forecasting system for South America. *Journal of Climate, 19*(15), 3704–3721. https://doi.org/10.1175/JCLI3801.1

Coelho, C. A. S., de Souza, D. C., Kubota, P. Y., et al. (2021). Evaluation of climate simulations produced with the Brazilian global atmospheric model version 1.2. *Climate Dynamics, 56*, 873–898. https://doi.org/10.1007/s00382-020-05508-8

CRC-SAS. (2023). *Descripción de la base de datos climáticos diarios y los controles de calidad implementados en el Centro Regional del Clima para el Sur de Sudamérica*. Reporte técnico CRC-SAS-2023-001 (spanish). https://sissa.crc-sas.org/wp-content/uploads/2023/11/Reporte_CRC-SASS_FINAL_calidad_19oct23.pdf

CRC-SAS. (2024). *Descripción de productos SISSA de precipitación y estado de la sequía derivados con datos satelitales (spanish)*. https://sissa.crc-sas.org/wp-content/uploads/2024/02/ReporteTecnicoCRC_ppinfosatelital.pdf

de Andrade, F. M., Coelho, C. A. S., & Cavalcanti, I. F. A. (2019). Global precipitation hindcast quality assessment of the Subseasonal to Seasonal (S2S) prediction project models. *Climate Dynamics, 52*(9–10), 5451–5475. https://doi.org/10.1007/s00382-018-4457-z

De Vera, A., Flieller, G., Chaer, R., & Terra, R. (2022). Evaluación de un sistema operativo de pronósticos hidrológicos por ensambles para la gestión del sistema eléctrico en Uruguay. *Ribagua, 9*(1), 10–24. https://doi.org/10.1080/23863781.2023.2213843

Diaz, L., et al. (2024). *Coproduction of a week-2 rainfall probabilistic prediction product to underpin small farmer activities in southeastern South America*. Submitted to Climate Services.

Escada, P., Coelho, C. A. S., Taddei, R., et al. (2021). Climate services in Brazil: Past, present, and future perspectives. *Climate Services, 24*, 100276. https://doi.org/10.1016/j.cliser.2021.100276

Funk, C., Peterson, P., Landsfeld, M., et al. (2015). The climate hazards infrared precipitation with stations—A new environmental record for monitoring extremes. *Scientific Data, 2*, 150066. https://doi.org/10.1038/sdata.2015.66

Han, E., Baethgen, W. E., Ines, A. V. M., et al. (2019). SIMAGRI: an agro-climate decision support tool. *Computers and Electronics in Agriculture, 161*, 241–251. https://doi.org/10.1016/j.compag.2018.06.034

Hernandez, V., Fossa Riglos, M. F., & Vera, C. S. (2022). Addressing climate services in South American Chaco region through a knowledge coproduction process. *Global Environmental Change, 72*, 102443. https://doi.org/10.1016/j.gloenvcha.2021.102443

Herrera, N. (2022). Pronóstico trimestral estadístico y estadístico-dinámico del Servicio Meteorológico Nacional. *Nota Técnica SMN*, 2022–2118. http://hdl.handle.net/20.500.12160/1789

Hewitt, C., Mason, S., & Walland, D. (2012). The global framework for climate services. *Nature Climate Change, 2*, 831–832. https://doi.org/10.1038/nclimate1745

Hewitt, C. D., et al. (2021). Recommendations for future research priorities for climate modeling and climate services. *Bulletin of the American Meteorological Society, 102*, E578–E588. https://doi.org/10.1175/BAMS-D-20-0103.1

Kolling Neto, A., Siqueira, V. A., Gama, C. H., et al. (2023). Advancing medium-range streamflow forecasting for large hydropower reservoirs in Brazil by means of continental-scale hydrological modeling. *Water, 15*, 1693. https://doi.org/10.3390/w15091693

Lemos, M. C., Kirchhoff, C. J., & Ramprasad, V. (2012). Narrowing the climate information usability gap. *Nature Climate Change, 2*, 789–794. https://doi.org/10.1038/nclimate1614

Loboguerrero Rodriguez, A. M., Hansen, J., Baethgen, W. E., & Martinez Baron, D. (2017). Climate services and insurance: Scaling climate smart agriculture. *Agriculture for Development, 30*, 31–34. https://hdl.handle.net/10568/81377

Maciel, F., Terra, R., & Chaer, R. (2015). Economic impact of considering El Niño-Southern Oscillation on the representation of streamflow in an electric system simulator. *International Journal of Climatology, 35*, 4094–4102. https://doi.org/10.1002/joc.4269

Mason, S. J. (2011). *Seasonal forecasting using the climate predictability tool (CPT)*. Science and Technology Infusion Climate Bulletin. NOAA's National Weather Service. 36th NOAA Annual Climate Diagnostics and Prediction Workshop. Fort Worth, TX, 3–6 October 2011. https://www.nws.noaa.gov/ost/climate/STIP/36CDPW/36cdpw-smason.pdf

Moraes, L. L. (2023). An impact-based forecast system developed for hydrometeorological hazards. *International Journal of Disaster Risk Reduction, 93*, 103803. https://doi.org/10.1016/j.ijdrr.2023.103803

NRC. (2010). *Assessment of intraseasonal to interannual climate prediction and predictability*. National Academies Press. https://doi.org/10.17226/12878

Osman, M., & Vera, C. S. (2017). Climate predictability and prediction skill on seasonal time scales over South America from CHFP models. *Climate Dynamics, 49*, 2365–2383. https://doi.org/10.1007/s00382-016-3444-5

Osman, M., Coelho, C. A. S., & Vera, C. S. (2021). Calibration and combination of seasonal precipitation forecasts over South America using ensemble regression. *Climate Dynamics, 57*, 2889–2904. https://doi.org/10.1007/s00382-021-05845-2

Petry, I., Mainardi Fan, F., Siqueira, V. A., et al. (2023). Seasonal streamflow forecasting in South America's largest rivers. *Journal of Hydrology: Regional Studies, 49*, 101487. https://doi.org/10.1016/j.ejrh.2023.101487

Ruscica, R. C., Sorensson, A. A., Diaz, L. B., et al. (2022). Evapotranspiration trends and variability in southeastern South America: The roles of land-cover change and precipitation variability. *International Journal of Climatology, 42*(4), 2019–2038. https://doi.org/10.1002/joc.7350

Terra, R., & Baethgen, W. E. (2019). Lessons learned in 25 years of informing sectoral decisions with probabilistic climate forecasts. In A. W. Robertson & F. Vitart (Eds.), *Sub-seasonal to seasonal prediction* (pp. 439–453). Elsevier.

Vaughan, C., Dessai, D., Hewitt, C., et al. (2017). Creating an enabling environment for investment in climate services: The case of Uruguay's National Agricultural Information System. *Climate Services, 8*, 62–71. https://doi.org/10.1016/j.cliser.2017.11.001

Vera, C. S. (2024). *Climate services in South America* (Oxford Research Encyclopedia of Climate Science). Oxford University Press. https://doi.org/10.1093/acrefore/9780190228620.013.963

Vera, C. S., Skansi, M. M., & Hidalgo, C. (2024). Co-designing strategies to manage drought related risks in southern South America. *WMO Bulletin, 73*(1), 15–17. https://library.wmo.int/idurl/4/68915

WMO. (2011). Climate knowledge for action: A global framework for climate services—Empowering the Most vulnerable. The report of the high-level taskforce for the global framework for climate services. In *World Meteorological Organization* (WMO no. 1065) https://library.wmo.int/idurl/4/35862

WMO. (2022). *Guidelines on the implementation of climate watches* (WMO No. 1299). World Meteorological Organization. https://library.wmo.int/idurl/4/58326

MIX
Papier aus verantwortungsvollen Quellen
Paper from responsible sources
FSC® C105338

If you have any concerns about our products,
you can contact us on
ProductSafety@springernature.com

In case Publisher is established outside the EU,
the EU authorized representative is:
Springer Nature Customer Service Center GmbH
Europaplatz 3, 69115 Heidelberg, Germany

Printed by Libri Plureos GmbH
in Hamburg, Germany